MEMOIRS
of the
American Mathematical Society

ISSN 0065-9266 (print)
ISSN 1947-6221 (online)

Number 1558

Construction of a Non-Gaussian and Rotation-Invariant Φ^4-Measure and Associated Flow on $\mathbb{R}^3$ Through Stochastic Quantization

Sergio Albeverio
Seiichiro Kusuoka

April 2025 • Volume 308 • Number 1558 (fifth of 6 numbers)

AMS AMERICAN MATHEMATICAL SOCIETY

Library of Congress Cataloging-in-Publication Data

Cataloging-in-Publication Data has been applied for by the AMS.
See http://www.loc.gov/publish/cip/.
DOI: https://doi.org/10.1090/memo/1558

Memoirs of the American Mathematical Society

This journal is devoted entirely to research in pure and applied mathematics.

Subscription information. Beginning in 2024, Memoirs will be published monthly through 2026. Memoirs is also accessible from www.ams.org/journals. The 2025 subscription begins with volume 305 and consists of twelve mailings, each containing one or more numbers. Individual subscription prices for 2025 are as follows. For electronic only: US$1044. For paper delivery: US$1197. Add US$65 for delivery within the United States; US$200 for surface delivery outside the United States. Upon request, subscribers to paper delivery of this journal are also entitled to receive electronic delivery. For information on institutional pricing, please visit `https://www.ams.org/publications/journals/subscriberinfo`. Subscription renewals are subject to late fees. See `www.ams.org/journal-faq` for more journal subscription information. Each number may be ordered separately; please specify number when ordering an individual number.

Back number information. For back issues see `www.ams.org/backvols`.

Subscriptions and orders should be addressed to the American Mathematical Society, P.O. Box 845904, Boston, MA 02284-5904 USA. *All orders must be accompanied by payment.* Other correspondence should be addressed to 201 Charles Street, Providence, RI 02904-2213 USA.

Memoirs of the American Mathematical Society (ISSN 0065-9266 (print); 1947-6221 (online)) is published bimonthly (each volume consisting usually of more than one number) by the American Mathematical Society at 201 Charles Street, Providence, RI 02904-2213 USA. Periodicals postage paid at Providence, RI. Postmaster: Send address changes to Memoirs, American Mathematical Society, 201 Charles Street, Providence, RI 02904-2213 USA.

This publication is indexed in *Mathematical Reviews®*, *Zentralblatt MATH*, *Science Citation Index®*, *Science Citation Index™-Expanded*, *ISI Alerting Services℠*, *SciSearch®*, *Research Alert®*, *CompuMath Citation Index®*, *Current Contents®/Physical, Chemical & Earth Sciences*.
This publication is archived in *Portico* and *CLOCKSS*.
Printed in the United States of America.

∞ The paper used in this book is acid-free and falls within the guidelines established to ensure permanence and durability.
Visit the AMS home page at `https://www.ams.org/`

10 9 8 7 6 5 4 3 2 1 30 29 28 27 26 25

To Solvejg, Mielikki, Seiko, Soichi, and Miori

Contents

Abstract

A new construction of non-Gaussian, rotation-invariant and reflection positive probability measures μ associated with the φ_3^4-model of quantum field theory is presented. Our construction uses a combination of semigroup methods, and methods of stochastic partial differential equations (SPDEs) for finding solutions and stationary measures of the natural stochastic quantization associated with the φ_3^4-model. Our starting point is a suitable approximation $\mu_{M,N}$ of the measure μ, which we intend to construct. $\mu_{M,N}$ is parametrized by an M-dependent space cut-off function $\rho_M : \mathbb{R}^3 \to \mathbb{R}$ and an N-dependent momentum cut-off function $\psi_N : \widehat{\mathbb{R}}^3 \cong \mathbb{R}^3 \to \mathbb{R}$, that act on the interaction term (nonlinear term and counterterms). The corresponding family of stochastic quantization equations yields solutions $(X_t^{M,N}, t \geq 0)$ that have $\mu_{M,N}$ as an invariant probability measure. By a combination of probabilistic and functional analytic methods for singular stochastic differential equations on negative-indices weighted Besov spaces (with rotation invariant weights) we prove the tightness of the family of continuous processes $(X_t^{M,N}, t \geq 0)_{M,N}$. Limit points in the sense of convergence in law exist, when both M and N diverge to $+\infty$. The limit processes $(X_t; t \geq 0)$ are continuous on the intersection of suitable Besov spaces and any limit point μ of the $\mu_{M,N}$ is a stationary measure of X. μ is shown to be a rotation-invariant and non-Gaussian probability measure and we provide results on its support. It is also proven that μ satisfies a further important property belonging to the family of axioms for Euclidean quantum fields, namely it is reflection positive.

Received by the editor March 5, 2021, and, in revised form, July 15, 2022, September 25, 2022, July 15, 2022, and September 25, 2022.

Article electronically published on April 8, 2025.

DOI: `https://doi.org/10.1090/memo/1558`

2020 *Mathematics Subject Classification.* Primary 81S20; Secondary 81T08, 60H15, 35Q40, 35R60, 35K58.

Key words and phrases. stochastic quantization, quantum field theory, singular SPDE, invariant measure, flow.

This work was supported by JSPS grants KAKENHI Grant numbers 25800054, 17K14204 and 21H00988.

CHAPTER 1

Introduction

1.1. Background

This paper is concerned with the construction of solutions of a certain stochastic partial differential equation (SPDE) of parabolic type, with singular coefficients, known as stochastic quantization equation (SQE) for the φ_3^4-model of relativistic quantum fields on the Euclidean 3-dimensional space-time $\mathbb{R}^3$. Particular attention is given to the construction of an invariant probability measure for the process associated with the SQE, that is stationary (i.e. invariant) with respect to the action induced on the state space by the Euclidean group (including reflections) of rigid transformations of $\mathbb{R}^3$. The property of Euclidean invariance is particularly important for assuring the relativistic invariance (i.e. invariance under the full Poincaré group) of the associated relativistic quantum fields. In this sense the construction of solutions of the SQE and their invariant measures provides an alternative construction of the ϕ_3^4-Euclidean and relativistic fields by other methods than those used in the constructive quantum field theory.

In order to better understand the motivations for our approach, let us shortly recall the origins of the problems that we are dealing with (for more details see also the references given below, and the introduction in our previous paper [**AK20**] and in the paper by Gubinelli and Hofmanová [**GH21**]). The most simple classical relativistic equation describing scalar waves is the linear wave equation

$$\square\varphi_{\rm cl}(t,\vec{x}) = -m_0^2\varphi_{\rm cl}(t,\vec{x}) \tag{1.1}$$

where $\varphi_{\rm cl}$ is a real valued function from the Minkowski space-time $M^d = \mathbb{R}\times\mathbb{R}^\sigma$, $\sigma\in\mathbb{N}\cup\{0\}$, $d := 1+\sigma$, $\square = \frac{\partial^2}{\partial t^2} - \triangle_{\vec{x}}$ being the d'Alembert operator, $(t,\vec{x})\in M^d$, with t standing for time and $\vec{x}$ for space. $m_0 > 0$ is a mass parameter. The modification of the equation (1.1) by some local nonlinear term $-\lambda V'(\varphi_{\rm cl}(t,\vec{x}))$, where $V\in C^1(\mathbb{R};\mathbb{R})$ and $\lambda > 0$ is a parameter, is the nonlinear Klein-Gordon equation of M^d

$$\square\varphi_{\rm cl} = -m_0^2\varphi_{\rm cl} - \lambda V'(\varphi_{\rm cl}). \tag{1.2}$$

To discuss the quantization of (1.2) it is useful to consider first the case that $\lambda = 0$ and $\sigma = 0$. In this case the quantization of (1.2) reduces to the quantization of the solution of the equation

$$\frac{d^2}{dt^2}\varphi_{\rm cl}(t) = -m_0^2\varphi_{\rm cl}(t).$$

This is just Newton's equation describing the evolution in time t of a classical harmonic oscillator. The quantized version $\varphi(t)$ of $\varphi_{\rm cl}(t)$ can be realized by identifying φ at $t=0$ with the multiplication operator by the coordinate in the complex Hilbert space $\mathcal{H} = L^2(\mathbb{R}, dx)$ and letting evolve $\varphi(0)$ by the unitary group generated by the self-adjoint Hamilton operator in $\mathcal{H}$ given on a dense domain $\mathcal{S}(\mathbb{R})$, e.g., by

$$H = -\frac{1}{2}\frac{d^2}{dx^2} + \frac{m_0^2}{2}x^2 - \frac{m_0}{2}.$$

One has on a dense domain $\varphi(t) = e^{-itH}\varphi(0)e^{itH}$ $(t \in \mathbb{R})$, where e^{itH} is the unitary group on $\mathcal{H}$ generated by H. Of particular interest for this case (but also for the wanted quantized version of (1.2)) is the study of the mean value of products of the operators φ taken at different times, the average being understood in the sense of scalar products taken in the Hilbert space. In fact it turns out that it is enough to look at the quantities

$$(\Omega, \varphi(t_1)\cdots\varphi(t_n)\Omega) \tag{1.3}$$

where $(\cdot,\cdot)$ is the scalar product in $\mathcal{H}$ and Ω is the (normalized) eigenfunction to the lowest eigenvalue, zero, of H (Ω is called "ground state" or "vacuum" state). A simple calculation shows that Ω is the function $x \mapsto \left(\frac{m_0}{\pi}\right)^{1/4} e^{-\frac{m_0}{2}x^2}$ in $\mathcal{H}$. (1.3) is a special case (for $\sigma = 0$ and $\lambda = 0$) of Wightman's functions that were found out to be of particular interest for the study of the quantization of (1.2) (and more general nonlinear Klein-Gordon equations in the quantum field theory, see, e.g., [**Jos65**], [**SW78**, Chapter 3-4], [**BLT75**] and [**Str13**, Chapter 2]).

For the construction of a corresponding structure for the case that $\sigma \geq 1$, the first problem we meet is the choice of Hilbert space. For this it is useful to recall (from non relativistic quantum mechanics) that for the harmonic oscillator the Hilbert space is $\mathcal{H} = L^2(\mathbb{R}, dx)$ and the quantized version $\varphi(0)$ of $\varphi_{\rm cl}(0)$ acts as multiplication by the coordinate function x. If we try to adapt this directly for quantum fields $\varphi(t, \vec{x})$, $\vec{x} \in \mathbb{R}^\sigma$, the operator in $\varphi(0, \vec{x})$ will be multiplication by a function of $\vec{x}$ that will belong to some infinite-dimensional space Σ, but then the measure structure (in analogy with the above dx) should be given by a "volume measure" on Σ. Such an approach meets difficulties, in fact it does not work if Σ is a Hilbert space (see, e.g. [**AM15**] and references therein for the absence of regular σ-additive measures on infinite-dimensional Hilbert spaces). To find out of this "impasse", it is useful to realize that $L^2(\mathbb{R}, dx)$ for the harmonic oscillator can be replaced by a suitable unitary equivalent space. E.g. the map sending $f \in L^2(\mathbb{R}, dx)$ into $\frac{f}{\sqrt{\rho}} \in L^2(\mathbb{R}, \mu_0)$, where μ_0 is the probability measure on the Borel subsets of $\mathbb{R}$ given by $\mu_0(dx) := \rho(x)dx$, with $\rho(x) := \left(\frac{m_0}{\pi}\right)^{1/2} e^{-m_0 x^2}$ $(x \in \mathbb{R})$, does the task. μ_0 is a centered Gaussian measure (with variance $(2m_0)^{-1}$), and $f \mapsto \frac{f}{\sqrt{\rho}}$ is a unitary map between $L^2(\mathbb{R}, dx)$ and $L^2(\mathbb{R}, \mu_0)$.

Note that by this map the Hamiltonian H is unitary equivalent to $H_{\mu_0} = -\frac{1}{2}\frac{d^2}{dx^2} + m_0 x \frac{d}{dx}$ acting on a dense domain in $L^2(\mathbb{R}, \mu_0)$. More precisely, this is the unique self-adjoint positive operator properly associated with the classical Dirichlet form $\mathcal{E}_{\mu_0}(u, v) = \frac{1}{2}\int_{\mathbb{R}} \frac{du}{dx}\frac{dv}{dx}\mu_0(dx)$ given by μ_0 (with u, v in a dense domain in $L^2(\mathbb{R}, \mu_0)$, see for the relations between Hamiltonians and classical Dirichlet forms, e.g. [**AHKS77**], [**AU15**], [**BK95a**, Chapter 1-2], [**BK95b**, Chapter 6]) and [**Sim05**, Chapter V, Section 16 and Chapter II, Section 4]. Note that the function identically equal to 1 in $L^2(\mathbb{R}, \mu_0)$ is an eigenfunction to the eigenvalue zero of H_{μ_0}. The stationary stochastic process properly associated to $\mathcal{E}_{\mu_0}$ (see [**FOT94**]) is an Ornstein-Uhlenbeck (velocity) process, with a covariance operator given by $\frac{e^{-m_0|t-s|}}{2m_0}$ $(s, t \in \mathbb{R})$, and this is the kernel of the operator $(-\frac{d^2}{dt^2} + m_0^2)^{-1/2}$ (see e.g. [**AKKR09**]). These observations are appropriate for an extension to the case $\sigma \geq 1$, that we now discuss. In the latter case μ_0 is replaced by the centered Gaussian measure $\mu_{0,\sigma}$ on $\mathcal{S}'(\mathbb{R}^\sigma)$ given uniquely on the basis of Minlos's theorem by its Fourier transform

$$\begin{aligned}\hat{\mu}_{0,\sigma}(f) &= \int_{\mathcal{S}'(\mathbb{R}^\sigma)} e^{i\langle f,\omega\rangle}\mu_0(d\omega)\\ &= \exp\left(-\frac{1}{2}\langle f,(-\triangle_\sigma+m_0^2)^{-1/2}f\rangle_{L^2(\mathbb{R}^\sigma)}\right)\end{aligned}$$

with $\triangle_\sigma$ the Laplacian in $L^2(\mathbb{R}^\sigma, dx)$ and $f \in \mathcal{S}(\mathbb{R}^\sigma)$. $\mu_{0,\sigma}$ is the ground-state measure for the quantum fields $\varphi(0,\vec{x})$ at time zero for a d $(=\sigma+1)$-dimensional space time, $(t,\vec{x}) \in \mathbb{R}\times\mathbb{R}^\sigma$. The Hilbert space in which they act would be $L^2(\mathcal{S}'(\mathbb{R}^\sigma), \mu_{0,\sigma})$ (in the case that $\sigma = 0$, $\mathcal{S}'(\mathbb{R}^\sigma)$ is replaced by $\mathbb{R}$ and $\mu_{0,\sigma}$ by μ_0 as above). Proceeding on the same line we can define a corresponding quantum Hamiltonian $H_{\mu_{0,\sigma}}$ given by $\mu_{0,\sigma}$ as the unique self-adjoint operator properly associated with the Dirichlet form $\mathcal{E}_{\mu_0}(u,v) = \frac{1}{2}\int \nabla u\cdot\nabla v d\mu_{0,\sigma}$ given by $\mu_{0,\sigma}$ (on its natural domain). The existence and the well-definedness of the Hamiltonian are proven in [**AHK77**] and [**AR89**]. $H_{\mu_{0,\sigma}}$ corresponds indeed to the classical Hamiltonian given (modulo an additive constant) by

$$\frac{1}{2}\int\left|\frac{\partial}{\partial t}\varphi_{\mathrm{cl}}(t,\vec{x})\Big|_{t=0}\right|^2 d\vec{x} + \frac{1}{2}\int\left|\nabla_{\vec{x}}\varphi_{\mathrm{cl}}(t,\vec{x})|_{t=0}\right|^2 d\vec{x} + \frac{m_0^2}{2}\int|\varphi(0,\vec{x})|^2 d\vec{x}.$$

(Incidentally $L^2(\mathcal{S}'(\mathbb{R}, \mu_{0,\sigma}))$ is isomorphic in a natural way, via the Friedrichs-Segal isomorphism, to the Fock space for the free time-zero quantum field $\varphi(0,\vec{x})$ (see e.g. [**BSZ92**, Chapter 1] and [**Sim74**, Chapter I])).

Note that the variance of $\varphi(0,\vec{x})$ is heuristically $(-\triangle_\sigma+m_0^2)^{-1/2}$ valuated at 0, which diverges, hence the time-zero quantum field "operator $\varphi(0,\vec{x})$" should be really understood as a $\mathcal{S}'(\mathbb{R}^\sigma)$-valued Gaussian random variable with distribution $\mu_{0,\sigma}$:

$$\varphi(f) := \langle f,\varphi\rangle, \quad f\in\mathcal{S}(\mathbb{R}^\sigma),\ \varphi\in\mathcal{S}'(\mathbb{R}^\sigma),$$

so that the first constructive approach of models given by (1.2) was developed in this frame work. The operator

$$\varphi(t,f) := e^{-itH_{\mu_{0,\sigma}}}\varphi(f)e^{itH_{\mu_{0,\sigma}}}$$

is then the time t-field operator acting in the Hilbert space $L^2(\mu_{0,\sigma})$, with spatial test-function f. To perturb $H_{\mu_{0,\sigma}}$ by an interaction term a suggestion comes again from the classical Hamiltonian corresponding to (1.2) that, in the case $\lambda \neq 0$, contains a term $\lambda\int_{\mathbb{R}^\sigma} V(\varphi_{\mathrm{cl}}(0,\vec{x}))d\vec{x}$. Due to the fact that the quantum time-zero field $\varphi(0,\vec{x})$ has to be understood in the distributional sense, an expression like "$\lambda\int_{\mathbb{R}^\sigma} V(\varphi(0,\vec{x}))d\vec{x}$", has only a heuristic meaning. For $\sigma = 1$, the standard way out is to "regularize and renormalize", substituting the above heuristic expression by "$\lambda\int_{\mathbb{R}^\sigma} : V(\varphi_\varepsilon(0,\vec{x})) : g(\vec{x})d\vec{x}$", where $:\cdot:$ is the Wick ordering (with respect to $\mu_{0,1}$ (see e.g. [**Sim74**, Chapter I])), $\varepsilon > 0$ means convolution with a smooth function of compact support in momentum space tending to 1 as $\varepsilon \downarrow 0$ ("ultraviolet cut-off") and g is a smooth function of compact support in space ("infrared cut-off"), which one then eventually let converge to 1. In this way one gets a perturbed Hamiltonian $H_{\varepsilon,g}$ as a well-defined self-adjoint operator on $L^2(\mu_{0,\sigma})$. (In the case $\sigma = 2$ modifications by "counterterms" (see below) are needed.) This is indeed the approach used in the first period of constructive quantum field theory (see e.g. [**BSZ92**, Chapter 8], [**GJ81**, Chapter 9] [**GJ68**], [**GJ70a**], [**GJ70c**], [**GJ70b**], [**Hep69**, Chapter IV], [**HK71**] and [**Sim74**, Chapter VII-VIII]; for the case that $\sigma = 0$ see also [**CR75**] and [**Str81**]). For a well-defined limit for $\varepsilon \downarrow 0$, $g \to 1$ for $\sigma = 1$ and for V of lower bounded polynomial type, a probabilistic lower bound for $H_{\varepsilon,g}$ found by E. Nelson played a crucial role (see [**Nel66**]; see also [**Nel81**] for Nelson's use of probabilistic ideas in quantum theory). This work was extended to other types of interaction terms V, namely an exponential or trigonometric term, for $\sigma = 1$ (see [**HK71**]). For the case where $\sigma = 2$ these Hamiltonian methods were extended to the case of V of a fourth power (the φ_3^4-model) in [**Gli68**]. But, major further developments, both for $\sigma = 1$ and $\sigma = 2$ came after "Euclidean methods" were introduced in the area of constructive quantum field theory.

Physicists like F. Dyson and G. C. Wick (in 1949 resp. 1954-56, see references in [**Mil15**]) already used Euclidean methods instrumentally, for going from the relativistic "Feynman propagator" associated with $(\Box - m_0^2)^{-1}$, appearing in "perturbative quantum field models" to the easier to handle Euclidean propagator $(-\triangle_d + m_0^2)^{-1}$, the passage being effectuating by formal analytic continuation from the relativistic time $t \in \mathbb{R}$ to a Euclidean time $\tilde{t} = -it$, by which $\Box = \frac{\partial^2}{\partial t^2} - \triangle_{\vec{x}}$ is mapped into $-\triangle_d = -\frac{\partial^2}{\partial \tilde{t}^2} - \triangle_{\vec{x}}$ ($\vec{x} \in \mathbb{R}^\sigma$, $\triangle_d$ being the Euclidean invariant Laplace operator on $\mathbb{R}^d$, $d = \sigma + 1$). Starting from 1956 an axiomatic framework for relativistic quantum field theory (extended from physicist's experience with the theory) was presented (see e.g. [**Wig56**], [**SW78**, Chapter 3], [**Jos65**, Chapter III], [**BLT75**], [**Str13**, Chapter 5] and reference therein) defined in terms of "correlation functions", called Wightman functions. These functions describe correlations expressed by mean values in a certain state, called "vacuum state", of n-fold products of quantum field operators at space-time x_i ($i = 1, 2, \dots, n$) in d-dimensional Minkowski space-time $\mathcal{M}^d$ (for any $n \in \mathbb{N}$). These are required to satisfy minimal properties, next to relativistic invariance, a spectrum condition for the generator of time evolution, a cluster property and a positivity condition (assuring a complex Hilbert space structure for the physical states). These properties were translated in physical terms by analytic continuation of the (time-ordered) Wightman functions in the time variables t_k to Euclidean time variables $\tilde{t}_k$ resulting in corresponding correlation functions, called Schwinger functions, of the Euclidean framework in work by J. Schwinger (1951 and 1958) and T. Nakano (1959), see [**Mil15**] (risp. [**Sch94**] for the relativistic case) for historical references.

A more mathematically oriented point of view on the Euclidean correlation functions was taken by K. Symanzik, who proposed Euclidean methods to construct relativistic models [**Sym69**] and [**Sym66**]. This program was taken up by E. Nelson in 1971, when he was lecturing in Princeton on mathematical themes in relation to [**Sym69**]. Nelson's work was followed by a series of publications (see [**Nel73d**], [**Nel73a**], [**Nel73b**] and [**Nel73c**]). The ground breaking observation was to look at the time-zero field measures, in particular $\mu_{0,\sigma}$ on $\mathcal{S}'(\mathbb{R}^\sigma)$, as time-zero marginal measures of Euclidean invariant measures $\mu_{0,\mathrm{E}}$ on $\mathcal{S}'(\mathbb{R}^d)$ with $d = \sigma + 1$ ($\sigma \geq 0$), $\mu_{0,\mathrm{E}}$ being by definition the probability measure on $\mathcal{S}'(\mathbb{R}^d)$ given by its Fourier transform

$$\widehat{\mu_{0,\mathrm{E}}}(f) = \int_{\mathcal{S}'(\mathbb{R}^d)} e^{i\langle f,\omega\rangle} \mu_{0,\mathrm{E}}(d\omega)$$
$$= \exp\left(-\frac{1}{2}(f, (-\triangle + m_0^2)^{-1} f)_{L^2(\mathbb{R}^d)}\right)$$

for any $f \in \mathcal{S}(\mathbb{R}^d)$, $\triangle := \triangle_d$ being the Laplacian on $\mathbb{R}^d$. The corresponding coordinate process $\varphi(h) := \langle h, \varphi\rangle$, $h \in \mathcal{S}(\mathbb{R}^d)$, $\varphi \in \mathcal{S}'(\mathbb{R}^d)$ is a Gaussian random variable on $\mathcal{S}'(\mathbb{R}^d)$ with mean zero and variance

$$\int \langle h_1, \varphi\rangle\langle h_2, \varphi\rangle \mu_{0,\mathrm{E}}(d\varphi) = \big\langle h_1, (-\triangle + m_0^2)^{-1} h_2\big\rangle_{L^2(\mathbb{R}^d)} \tag{1.4}$$

i.e. φ is a Gaussian random process (indexed by $\mathcal{S}(\mathbb{R}^d)$). The right-hand side can be looked upon as a scalar product in the Sobolev space $H^{-1}(\mathbb{R}^d)$ and one can naturally look upon φ as a random process supported by $H^{-1}(\mathbb{R}^d)$. The full Euclidean group $E(\mathbb{R}^d)$ on $\mathbb{R}^d$ (generated by translations, rotations and reflections) leaves $\mu_{0,\mathrm{E}}$ invariant, in the sense that, for any $\beta \in E(\mathbb{R}^d)$, one has $\mu(T_\beta(A)) = \mu(A)$ for any Borel subset in $H^{-1}(\mathbb{R}^d)$, T_β being the pointwise map on $H^{-1}(\mathbb{R}^d)$ into itself given by

$$\varphi(h) \circ T_\beta = \varphi(h \circ \beta), \quad h \in H^{-1}(\mathbb{R}^d).$$

The fact mentioned above that the measure $\mu_{0,\sigma}$ can be looked upon as time-zero marginal is due to the reflection symmetry and strict Markov property of $\mu_{0,\mathrm{E}}$ with respect to (the σ-algebra associated with) the hyperplane $\Pi_0 := \{(x^1, \vec{x}); x^1 = 0, \vec{x} \in \mathbb{R}^\sigma\}$ in $\mathbb{R}^d$, $d = \sigma + 1$. To see this, $\mu_{0,\mathrm{E}}$ being Gaussian, it is enough to examine its covariance, given by (1.4). This has a special structure, best seen by observing that (because of the support properties of $\mu_{0,\mathrm{E}}$) the right-hand side of (1.4) is also finite if $h_i \in \mathcal{S}(\mathbb{R}^d)$ $(i = 1, 2)$ are replaced by the generalized functions of the form $\delta_{t_i}(x_i^1)\chi_i(\vec{x})$ ($x_i^1 \in \mathbb{R}$, $\vec{x}_i \in \mathbb{R}^\sigma$, $t_i \in \mathbb{R}$ and $\chi_i \in \mathcal{S}(\mathbb{R}^\sigma)$, $i = 1, 2$), and one has, by Fourier transform and an application of the "method of residua" for performing an integration over the first coordinate in $\widehat{\mathbb{R}}^d$ (the dual of $\mathbb{R}^d$, $\widehat{\cdot}$ used for Fourier transform):

$$\big\langle h_1, (-\triangle + m_0^2)^{-1} h_2\big\rangle_{L^2(\mathbb{R}^d)} = \int_{\widehat{\mathbb{R}}^\sigma} \overline{\widehat{\chi_1}}(\vec{p}) \frac{e^{-\sqrt{\vec{p}^2 + m_0^2}|t_2 - t_1|}}{\sqrt{\vec{p}^2 + m_0^2}} \widehat{\chi_2}(\vec{p}) d\vec{p}.$$

Note that this is positive for $\chi_1 = \chi_2$, hence one has on the right-hand side a scalar product in $L^2(\widehat{\mathbb{R}}^\sigma)$. From this it is not difficult to deduce that the sharp time random field $\varphi(t, \vec{x})$ looked upon as a stochastic process in $t \in \mathbb{R}$ with state space $\mathcal{S}'(\mathbb{R}^\sigma)$ is a symmetric Markov process with continuous paths and invariant measure $\mu_{0,\sigma}$.

Note that $\mu_{0,\sigma}$ has covariance $(-\triangle_\sigma + m_0^2)^{-1/2}$. Its transition semigroup can be identified with the one generated by the unique self-adjoint operator $H_{\mu_{0,\sigma}}$ in $L^2(\mu_{0,\sigma})$ associated with the classical Dirichlet form given by $\mu_{0,\sigma}$ (see e.g. [**Alb03**] and [**AR89**]). The unitary group generated by $H_{0,\sigma}$ has then the interpretation of time-translation group for the free relativistic quantum field over Minkowski space-time $\mathcal{M}^d$ (corresponding to the Euclidean space $\mathbb{R}^d$, $d = 1 + \sigma$). This relies on the explicit form of $(-\triangle + m_0^2)^{-1}$ i.e. the inverse of a local second-order Euclidean invariant operator (see [**Nel73b**], [**Sim74**, Chapter III], [**GRS75**], [**AHK77**], [**Röc85**]). To introduce interactions, the first idea inspired by classical field theory (or the case $\sigma = 0$ we discussed above) is to consider a Feynman-Kac or Gibbs type measure

$$\mu_{\varepsilon,g}(d\varphi) := Z_{\varepsilon,g}^{-1} e^{-\int_{\mathbb{R}^d} V(\varphi_\varepsilon(x))g(x)dx} \mu_{0,\mathrm{E}}(d\varphi) \tag{1.5}$$

where $\varepsilon > 0$ is a regularization ("ultraviolet cut-off") and g a smooth function with compact support in $\mathbb{R}^d$ ("infrared cut-off"), $Z_{\varepsilon,g}$ is a normalization constant. The passage to the limit $\varepsilon \downarrow 0$ and $g \to 1$ (besides requiring renormalization) does on one hand at least heuristically guarantee locality of the interaction ($\varepsilon \downarrow 0$) and Euclidean invariance ($g \to 1$), but it does not guarantee that the strict Markov property of the new measure (in case it can be obtained by such a limit) holds, and that therefore a Hamiltonian operator exists. In fact the proof of the strict Markov property for the exponential and polynomial interaction models , even for $\sigma = 1$, was achieved only some years later in [**AHK74**], [**AHKZ89**], [**Zeg84**], [**Gie83**] and references therein and also e.g. [**Röc88**] (for $\sigma = 2$ it is still open). At this point a basic observation by Osterwalder and Schrader came in.

To describe it let us first recall that from the physical point of view the basic goal is to obtain in the limit for $\varepsilon \downarrow 0$ and $g \to 1$ a unitary representation of the Euclidean group acting on a Hilbert space $\mathcal{H}$ associated with the limit μ of $\mu_{\varepsilon,g}$ and the corresponding limit fields $\varphi(x)$ ($x \in \mathbb{R}^d$). The natural candidate for $\mathcal{H}$ is either $L^2(\mathcal{S}'(\mathbb{R}^d), \mu)$, μ being the limit of $\mu_{\varepsilon,g}$ for $\varepsilon \downarrow 0$ and $g \to 1$, or $L^2(\mathcal{S}'(\mathbb{R}^d), \mu \restriction \Sigma_0)$, $\mu \restriction \Sigma_0$ being the restriction of the probability measure μ to the sub-σ-algebra in $\mathcal{S}'(\mathbb{R}^d)$ generated by the time zero field. μ should have moments, Schwinger functions, that are then Euclidean invariant. In addition the inverse operation of analytic continuation (corresponding to the inverse of the one we described before, passing from Wightman to Schwinger functions) should lead to relativistic invariant functions (Wightman functions). Nelson [**Nel66**]–[**Nel73d**] (see also [**Frö74**] and [**Sim74**, Chapter II]) realized that such an analytic continuation is possible if the limit μ satisfies a set of "Euclidean axioms", including the strict Markov property, and then the Wightman functions so obtained would satisfy the Wightman axioms. Osterwalder and Schrader's basic observation was that a weaker Euclidean axioms system for a set of Schwinger functions also leads to the satisfaction of the Wightman axioms. There are various versions of these axioms, see [**OS73**], [**OS75**], [**Ost82**], [**Ost73**], [**Frö74**], [**Sim74**, Chapter II and IV], [**GJ81**, Chapter 6 and 12] (see also [**Heg74**] and [**Kle78**] for a special case, intermediate between Nelson's axioms and Osterwalder-Schrader axioms). For successive discussions of the axioms including their equivalence see [**Gla74**] and [**Zin95**] and references therein. Osterwalder and Schrader underlined a "reflection positivity" axiom as a weaker requirement than the strict Markov property that can be more easily verified in models, being formulated through inequalities that are stable under passage to the limit.

Let us formulate reflection positivity in terms of the Fourier transform of a measure μ on $\mathcal{S}'(\mathbb{R}^d)$. μ is said to be reflection positive if the following two requirements are satisfied:

(i) μ is reflection symmetric, i.e. for any Borel subset A in $\mathcal{S}'(\mathbb{R}^d)$, we have $\mu(A) = \mu(T_\theta(A))$, where $T_\theta(A)$ is the subset of $\mathcal{S}'(\mathbb{R}^d)$ obtained from A by the transformation T_θ induced in $\mathcal{S}'(\mathbb{R}^d)$ by the underlying reflection $(x^1, \vec{x}) \mapsto (-x^1, \vec{x})$ in $\mathbb{R}^d$.

(ii) it satisfies the property that

$$\int_{\mathcal{S}'(\mathbb{R}^d)} e^{i\langle\varphi,{}^\theta f_j - f_k\rangle}\mu(d\varphi)$$

is a positive definite matrix, for any $N \in \mathbb{N}$ and $\{f_j \in \mathcal{S}(\mathbb{R}^d); j = 1,2,\ldots,N\}$ with supports in $\mathbb{R}^d_+ = \{(x^1,\vec{x}) \in \mathbb{R}^d; x^1 > 0\}$.

By (ii) one has a scalar product on the subspace Σ_+ in $L^2(\mathcal{S}'(\mathbb{R}^d),\mu)$ spanned by the set of functions $\{\mathcal{S}'(\mathbb{R}^d) \ni \varphi \mapsto e^{i\langle\varphi,h_k\rangle}; h_k \in \mathcal{S}(\mathbb{R}^d), \operatorname{supp} h_k \subset \mathbb{R}^d_+\}$. An important result in [**OS73**], [**OS75**] and [**Ost82**] (see also [**Frö74**]) is that reflection positivity yields already a contraction semigroup $\{P_t; t \geq 0\}$ in a Hilbert space $\mathcal{H}$ associated with μ, whose generator will also generate a unitary time translation group in a relativistic framework which we are going to describe shortly.

The passage from Euclidean (correlation) functions having the reflection positivity (RP) property to corresponding relativistic (correlation) functions in essence goes as follows. Let us assume, as before, that the Euclidean functions are moments of a given probability measure on $\mathcal{S}'(\mathbb{R}^d)$, assumed to be reflection positive and Euclidean invariant. Let $x = (x^1,\vec{x})$ ($x^1 \in \mathbb{R}$ and $\vec{x} \in \mathbb{R}^\sigma$), and let $U(t)$ ($t \in \mathbb{R}$) be the unitary group representing translations in direction of $x^1 = t$ (thought as "time") in $L^2(\mu) := L^2(\mathcal{S}'(\mathbb{R}^d),\mu)$. By the RP condition, for $F \in \Sigma_+$ we have that $\int {}^\theta F F d\mu \geq 0$. Let $\mathcal{N}$ be the corresponding null space (i.e. $\mathcal{N} := \{F \in \Sigma_+ |\ \int {}^\theta F F d\mu = 0\}$). Let $\mathcal{H}$ be the completion of $\Sigma_+|\mathcal{N}$ in the positive scalar product given by $\int {}^\theta F F d\mu$. $\mathcal{H}$ is the candidate for the Hilbert space for a corresponding relativistic setting. The canonical projection j satisfies $\|j(F)\|_{\mathcal{H}} \leq \|F\|_{L^2(\mu)}$, as easily seen. Let us consider on $j\mathcal{H}$ the 1-parameter family P_t defined for $t \geq 0$ by $P_t j(F) := j(U(t)F)$ ($F \in \Sigma+$). Since $U(t)\mathcal{N} \subset \mathcal{N}$, P_t is well-defined and extends uniquely to $\mathcal{H}$, it also easy to see that P_t is a strongly continuous semigroup on $\mathcal{H}$ and moreover, as shown in [**OS73**], [**OS75**] and [**Kle78**, Lemma 1.8], P_t is a contraction on $\mathcal{H}$. The generator H of P_t will be identified with the relativistic Hamiltonian. Let $\Omega = j(\mathbf{1})$, where $\mathbf{1}$ is the identity function in $L^2(\mu)$. Then $\|\Omega\|_{\mathcal{H}} = 1$ and $P_t\Omega = \Omega$ for all $t \geq 0$. Ω is going to be interpreted as relativistic vacuum. In fact from the above it is not difficult to see that $f_i \in \Sigma_0$ ($i = 1,2,\ldots,n$) one has, for $t_1 \leq t_2 \leq \cdots \leq t_n$:

$$\left(\Omega, \tilde{f}_1 P_{t_2-t_1}\tilde{f}_2 \cdots P_{t_n-t_{n-1}}\tilde{f}_n\Omega\right)_{\mathcal{H}} = \int f_{t_1} f_{t_2} \cdots f_{t_n} d\mu,$$

where $\tilde{f}_i j(F) = j(f_i F)$ ($F \in \Sigma_+$) and $f_{t_i} := U(t_i) f_i$ ($f_i \in \Sigma_0$). Under the assumption that the moments

$$S_n(f_1, f_2, \ldots, f_n) := \int \langle\varphi, f_1\rangle\langle\varphi, f_2\rangle \cdots \langle\varphi, f_n\rangle d\mu$$

of μ exists even when $f_i \in \mathcal{S}(\mathbb{R}^d)$ are replaced by $f_i = \delta_{t_i} \times k_i$ ($i = 1,2,\ldots,n$, $t_n > t_{n-1} > \cdots > t_1$, $k_i \in \mathcal{S}(\mathbb{R}^\sigma)$) one gets under some conditional regularity assumption ([**Frö74**, Axiom C]) that the Schwinger functions $S_n((t_1,\vec{x}_1),(t_2,\vec{x}_2),\ldots,(t_n,\vec{x}_n))$ coincide with the Wightman functions $W_n((it_1,\vec{x}_1),(it_2,\vec{x}_2),\ldots,(it_n,\vec{x}_n))$ at the Euclidean points $t_j \in \mathbb{R}$, $\vec{x}_j \in \mathbb{R}^\sigma$ with $(t_j,\vec{x}_j) = (t_l,\vec{x}_l)$ for $j \neq l$, the Wightman functions satisfying all Wightman axioms (with the possible exception of uniqueness of the vacuum) (see [**Frö74**], [**OS73**], [**OS75**], [**Sim74**, Theorem II.12 and II.13], [**Sum12**] and [**GJ81**, Theorem 12.1.1]).

Let us now see how this program is concretely implemented in the models we were discussing. We start by looking again at the regularized (through ultraviolet and space cut-off) probability measure μ_ε given by (1.5). The problem is now to find interpretations of ε and g such that the limit $\varepsilon \downarrow 0$ and $g \to 1$ of the Schwinger functions as moments of $\mu_{\varepsilon,g}$ exists and satisfies the axioms for Euclidean Schwinger functions. To this "minimal program" one should add other properties making the set of Wightman functions interesting also from the point of view of the physical interpretation (e.g. the limit measure should not be Gaussian or infinitely divisible, and a nontrivial scattering operator should exist, see e.g. [**Jos65**, Section VI.7]). For scattering theory and its background see [**Str13**, Section 2.6]. This program has succeeded in the cases $d = 2$ and $d = 3$, for some choices of cut-off and interaction terms V. The most complete picture is of course for the simpler case $d = 2$ and we shall first mention some results for this case, limiting ourselves to mainly give some indicative references. Here V is taken to be a lower bounded polynomial or a superposition of exponential or trigonometric terms, and V is replaced by a corresponding Wick ordered expression. In all these cases existence, and in some cases uniqueness, of Euclidean measures satisfying all Euclidean axioms and additional properties, have been shown, see e.g. [**GJ81**, Chapter 9-12], [**Sim74**, Chapter VIII and IX] and [**Frö74**] for the polynomial case (called $P(\varphi)_2$-model). The dependence of the choice of cut-offs has been discussed e.g. in [**GRS75**], [**FS77**] and [**AL04**]; for uniqueness results on the limit measure see [**Frö74**], [**Gue72**], [**GRS72**], [**GRS75**], [**AHKZ89**], [**ABR99**], [**AKR97**] and [**AR89**]. Also results on the structure of Gibbs states have been obtained in [**DM73**], [**AR89**] and [**GRS75**], together with other results, fully exploiting and adaptation of classical statistical mechanical methods, see also e.g. [**Nel73d**], [**GRS75**] and [**LN74**]. For results on the C^∞ character in λ for $\lambda \in (0, \delta]$ $(\delta > 0)$ of the Schwinger functions associated with the limit measure μ, see [**Dim74**], [**Dim77**] and [**Sum80**] (the derivatives being those indicated by perturbation theory). Asymptotic Borel summable expansion of Schwinger functions in λ for $V(y) = y^4$ has been established in [**EMS75**], as well as for the mass gap see [**EE79**], scattering was discussed in [**SZ76**] and [**DE77**]. For Borel summability in the case of the S-matrix see [**OS76**], [**EEF77**]. For results concerning semiclassical expansion of the Schwinger functions see [**Eck76**], and also e.g. [**Don81**]. Other results are related to the probabilistic structure of μ looking at its Markov character, associated Dirichlet forms and associated diffusion processes, see [**Nel73d**], [**Nel75**], [**GR73**], [**AHK77**], [**Röc85**], [**AGR01**], [**ARY17**], [**Alb03**] and [**MR92**, Chapter IV, Section 4b]. Other results concern the exponential (Høegh-Krohn or Liouville model) see [**HK71**], [**AHK74**], [**Sim74**, Section V.6], [**FS76**], [**FP77**], [**AHKZ89**] and [**AGHK79**]. For trigonometric models [**AHK73**], [**FS76**] and [**BGN82**] (see also for hyperbolic case, e.g. [**AHR96**] and [**AHR01**]). A structural result is concerning the measures μ for all such models for $d = 2$ is that μ can be identified as a Hida distribution (see [**AHP$^+$89a**], [**AHP$^+$89b**], [**ARY17**] and [**HKPS93**]; see also [**AFHKL86**] for a nonstandard analysis approach).

The case of special interest for the present paper is for $d = 3$, $V(y) = y^4$ (the only known nontrivial scalar model for which all axioms for $d = 3$ have been established). In this case besides changing V by Wick ordering further modifications by counterterms have been used (see (1.6) below in the context of the Euclidean method).

The first work on the construction of the φ^4_3-model is due to James Glimm [**Gli68**], who proved by Hamiltonian methods in the relativistic case with a space cut-off (i.e. in a bounded region for the space variable in $\mathbb{R}^2$) the existence of a Hilbert space and a densely defined symmetric Hamiltonian, the Hilbert space being different from the original Hilbert-Fock space for the interaction free case. Other Hamiltonian methods were used in the study of canonical commutation relations for the φ_3^4-model in a bounded spatial region in [**EO71**], where independence (in the sense of unitary equivalence) of the representations with respect to the space cut-off has been proven. For other work on the φ_3^4-model by Hamiltonian methods see also [**Hep69**, Chapter IV]. The non-locally Fock property of the (time zero) canonical commutation relation is proven in [**Fab70**] (see also [**Hep69**, Chapter I]). This was shown later in [**AL07**] to imply the singularity of the restriction of the Euclidean φ_3^4-measure to the one for the time-zero fields in a relative to the Euclidean free field bounded region measure (see also below for a recent paper by Barashkov and Gubinelli [**BG21**] for the singularity of the non-restricted Euclidean φ_3^4-measure with respect to the free field measure in a bounded Euclidean region). Ultraviolet stability and the positivity of the Hamiltonian (after removal of the ultraviolet cut-off, in a bounded Euclidean space-time volume) was proven in [**GJ73**] (see also [**GJ72**]). For other early works on renormalization for the φ_3^4-models see [**Fed71**] (and [**Bat99**, Chapter 5]). The usefulness of the introduction of Euclidean methods had already been exemplified in the study of the $P(\varphi)_2$-model in the whole $\mathbb{R}^2$-space, see [**Gue72**], [**GRS75**] and [**GJS74**]. To discuss Euclidean methods in the case of the φ_3^4-model it is useful to look back i.e. the (1.5) with $V(\varphi(x))$ with $V(y) = y^4$ $(y \in \mathbb{R})$ replaced by $V_\varepsilon((\varphi_\varepsilon(x)))$ with φ_ε and ultraviolet regularized φ, e.g. $\varphi_\varepsilon = \chi_\varepsilon * \varphi$, with χ_ε $(\varepsilon > 0)$ being an ε-approximation of the δ_0-measure on $\mathbb{R}^3$ i.e. $\chi_\varepsilon = \varepsilon^{-3} h(\varepsilon^{-1} \cdot)$ $(h \geq 0,\ \int_{\mathbb{R}^3} h(y)dy = 1,\ h \in C_0^\infty(\mathbb{R}^3))$ with

$$\lambda V_\varepsilon(\varphi_\varepsilon(x)) = \lambda \varphi_\varepsilon(x)^4 + a_\varepsilon(\lambda)\varphi_\varepsilon(x)^2$$

with $\lambda \geq 0$, $a_\varepsilon(\lambda) = -\alpha \frac{\lambda}{2} \varepsilon^{-1} - \beta \lambda^2 \log \varepsilon + \sigma$ (see [**GJ81**, Section 23.1]). Let then

$$\mu_{g,\varepsilon}(d\varphi) = Z_{g,\varepsilon}^{-1} e^{-\lambda \int_{\mathbb{R}^3} V_\varepsilon(\varphi_\varepsilon(x)) g(x) dx} \mu_0(d\varphi) \tag{1.6}$$

For $\mathbb{R}^3$ replaced by a torus $\mathbb{T}^3$ in $\mathbb{R}^3$, the existence of a limit μ_g of $\mu_{g,\varepsilon}$ for $\varepsilon \downarrow 0$ was shown in [**Fel74**]. The existence of a limit μ on $\mathbb{R}^3$ for $g \to 1$ and $\varepsilon \downarrow 0$ for σ large (equivalently, λ small enough: "small or weak coupling case") and the proof that all Wightman axioms are satisfied were given in [**FO76**] and [**MS76**] both using a "cluster expansion" method (inspired by classical statistical mechanics). Let us observe that it is on the basis of these particular results in the weak coupling case that in reality one can legitimately speak of "the φ_3^4-model" or "the φ_3^4-Euclidean measure". Other constructions reach quicker the goal of showing that there is a limit point for the family $(\mu_{\varepsilon,g})_{\varepsilon,g}$, but they have difficulties either with full Euclidean invariance (see [**BFS83**]; and also [**Par77**] and [**Par76**]) or reflection positivity (see [**AY02**] and [**AY09**]).

See also for other methods of construction of the φ_3^4-measure the approach using wavelet renormalization group [**Bat99**]. For other partly alternative constructions of the φ_3^4-Euclidean measure and corresponding Schwinger function, see also e.g. [**BCG$^+$78**] (who use a probabilistic approach); [**Bał83**] and [**Dim14**] (for a "block average approach"); [**GK86**] and [**BDH95**] (for methods by renormalization groups) (see also the very nice introduction of [**GH21**]).

For the weak coupling case the simplest and most elegant method to obtain limit points from a lattice versions of the φ_3^4-model is the work [**BFS83**] (see also [**FFS92**, Chapter 15]) where the important skeleton inequalities are proven. So one gets limit measures that are translation invariant and reflection positive (but not necessarily rotation invariant), at least for M sufficiently large (depending on λ) (weak coupling limit) (see [**BFS83**, Theorem 1.1]). For the full Euclidean invariance on this weak coupling limit an identification with the construction by cluster expansion methods is needed to be applied, which is obtained in the cases of some approximations (see e.g. [**FO76**], [**MS77**] for the construction by cluster expansion methods). Also for this weak coupling case the existence of a mass gap was shown and the first study of the particle structure was provided [**Bur**], [**SS76**] and [**BFS83**]. The particle structure mathematically means the description of the point spectrum and the continuous spectrum, and physically it explains stable states for particles. The asymptotic character of the perturbation expansion in powers of λ for the Schwinger functions associated with the Euclidean φ_3^4-measure was shown on the basis of the important skeleton inequalities in [**BF84**], and Borel summability for this expansion was established by a phase cell expansion in [**MS77**]. The fact that there is an isolated one particle state in the spectrum of the Hamiltonian is proven in [**Bur77**] and this permitted to apply to this model Haag-Ruelle methods for scattering theory (for the general Haag-Ruelle scattering theory see [**Jos65**]). Here, the existence of one particle state mathematically means that the infimum of the spectrum strictly greater than 0 is an isolated point in the spectrum. The asymptotic character of the scattering matrix elements was later shown in [**Con77**]. That the time ordered Wightman functions and the S-matrix elements are C^∞ in λ for $\lambda \in [0, \delta]$ ($\delta > 0$) was proven in [**EE79**]. In [**FRa77**] relativistic equations of motion for the model were discussed.

Results for large values of λ have also been established in particular the existence of at least two vacuum vectors, see [**FSS76**]. Concerning the structure of the space of Gibbs measures very little seems to be known. Known facts are e.g. the failure of Nelson's symmetry and the vacuum overlap (see [**SS76**]), and the mutual singularity for different λs (see [**FO76**]).

There is a study of solutions of the particular singular SPDEs (stochastic partial differential equations) appearing in quantum field theory under the name of stochastic quantization equations (SQE). The origin of the study of SQEs goes back to a very original approach by the physicists Giorgio Parisi and Yongshi Wu [**PW81**], who associated to a candidate classical evolution equation like the one, that we have in (1.2), a stochastic differential equation in such a way that the invariant measure for such an equation is precisely a relevant measure of interest connected with the original equation. The advantage of this procedure is that the solutions of the associated stochastic differential equation can in principle be used to make a Monte-Carlo type simulation for computing quantities associated to the measure of interest. In recent years, for the study of such singular SPDEs, new methods have been initiated particularly by Martin Hairer (with his theory of regularity structure) and Massimiliano Gubinelli (with his approach to SSPDEs through a theory of paracontrolled distributions). These methods remarkably work for SQEs, and now the study of SQEs is developing rapidly. In the case of (1.2), keeping in mind that the solution of it, as we discussed above, passed through the construction of the Euclidean measure μ indicated below by (1.8), the stochastic quantization equation takes the form of the singular SPDE

$$dX_\tau = \left[(\triangle - m_0^2)X_\tau - \lambda V'(X_\tau)\right] d\tau + dW_\tau, \quad \tau \geq 0, \tag{1.7}$$

where τ is the additional "computer time", X_τ is a process in τ depending on the variable $x \in \mathbb{R}^d$, $\triangle$ being the Laplacian in $\mathbb{R}^d$. dW_τ is a Gaussian white noise in τ and in x. In fact, heuristically the form of (1.7) suggests an invariant measure precisely of the form of the Euclidean measure μ given as

$$\mu(d\varphi) = \text{``}Z^{-1} e^{-\lambda \int_{\mathbb{R}^d} V(\varphi(x))dx} \mu_0(d\varphi)\text{''}, \tag{1.8}$$

μ_0 being Nelson's free field measure (indeed an invariant measure of (1.7) for the case $\lambda = 0$, having mean 0 and covariance suggested by the linear drift term in (1.7); the case $\lambda > 0$ can be mentally related to a Feynman-Kac type term perturbing μ_0, associated with λV). The program has been implemented for $d = 1$, 2 and 3. In the case $d = 1$ let us mention the work by T. Funaki on a bounded interval of $\mathbb{R}$ [**Fun83**], R. Marcus on $\mathbb{R}$ [**Mar79**], and especially Koichiro Iwata [**Iwa87a**] (much in the spirit of later work for the case $d = 2$). The latter studies $C(\mathbb{R}, \mathbb{R}^n)$-valued processes with multiplicative noise and a rather general nonlinear drift term; in the case where the noise is additive and the drift is given by a gradient field $-\lambda V'$ one has (1.7) (with V the sum of a convex C^1-function and a C^2-function with compact support, e.g. a "double-well potential"), see also [**Iwa87b**]. Questions of uniqueness of Dirichlet operators associated with the $P(\varphi)_1$-measure are discussed in [**KR07**] and the strong uniqueness is proven (on the full space $\mathbb{R}$). For previous work on such uniqueness questions see e.g. [**AR91**], [**Ebe99**], [**AKR12**], [**Shi95**], [**Tak85**] (who proved for the first time Markov uniqueness for models in $d = 1$). For uniqueness results on Gibbs measures that are associated with (1.7) for $d = 1$, see also [**BL03**]. See also the survey [**AMR15**] and references therein. For the case $d = 2$ the first results were obtained by replacing (1.7) by a more regular noise and modifying accordingly the drift term in order to maintain the same invariant Euclidean measure (we shall henceforth call the equation (1.7) with such modification as "SQE_ε equation", ε being a parameter in the modified noise such that for $\varepsilon \downarrow 0$ this noise becomes Gaussian white noise on $\mathbb{R}^2$). The first mathematical results on $(\text{SQE})_\varepsilon$ were obtained by G. Jona-Lasinio and S. Mitter [**JLM85**]. Weak probabilistic solutions are then obtained for SQE_ε for suitable initial conditions, both on the 2-dimensional torus $\mathbb{T}^2$ and on $\mathbb{R}^2$. Much further work was then done by various authors, see e.g. references in [**Alb03**], and Markov uniqueness for SQE or SQE_ε of the generator was shown in [**Tak85**]. Let us mention in particular the work [**DPT00**] for the SQE_ε on $\mathbb{T}^2$, where the generator of the Markov semigroup associated with SQE_ε is shown to be essentially self-adjoint on a natural domain with respect to the Euclidean invariant $P(\varphi)_2$-measure μ on $\mathbb{T}^2$. For the case $d = 2$ and $\varepsilon = 0$, the SQE on $\mathbb{T}^2$ has been solved strongly by probabilistic methods in [**DPD03**]. This paper introduced the by now commonly called "Da Prato-Debussche method" of splitting the solution X_τ as $X_\tau - Z_\tau + Z_\tau$, $(Z_\tau, \tau \geq 0)$ being the Ornstein-Uhlenbeck process associated with the linear part of (1.7) (modified by counterterms, in correspondence to those in $\mu_{\varepsilon,g}$) and exploiting then the known property of Wick powers of Z_τ. These authors also obtained the ergodicity of the solution process for μ-almost every initial data. In $\mathbb{R}^2$ the SQE has been discussed first by the method of Dirichlet forms in [**BCM88**], [**Mit91**], [**AR89**] and [**AR90**] for quasi-every initial conditions. In [**MW17c**] well-posedness of SQE for $d = 2$ in negative Besov spaces was shown both on $\mathbb{T}^2$ and $\mathbb{R}^2$. The complex of questions on uniqueness of Markov processes associated with the φ^4_2- and $P(\varphi)_2$-models and the one on uniqueness of the invariant measure have been further discussed in a series of papers. In [**RZZ17a**] ergodicity of the process associated with the φ_2^4-model on $\mathbb{T}^2$ is established as the extremality of the invariant measure (in the set of all L-symmetrizing measures, L being the corresponding generator). In [**RZZ17b**] restricted Markov uniqueness for the $P(\varphi)_2$-model both on $\mathbb{T}^2$ and $\mathbb{R}^2$, as well as ergodicity have been proven. For these results the authors use, in addition to methods of the theory of Dirichlet forms, results in [**MW17c**] on the construction of strong solutions of the SQE for the $P(\varphi)_2$-model in certain Besov spaces with negative indices. In [**HM18b**] the strong Feller property of the semigroup associated to the SQE for the $P(\varphi)_2$-model on $\mathbb{T}^2$ was proven (as related to their proof of the corresponding property for the φ_3^4-model on $\mathbb{T}^3$ (and the KPZ equation on $\mathbb{R} \times \mathbb{T}$)). Related results have been established in [**TW20**] where the authors showed in particular that the semigroup associated to the φ_2^4-model (on $\mathbb{T}^2$) maps bounded Borel functions into α-Hölder continuous functions for some $\alpha \in (0, 1)$. They also showed exponential speed of approximation to the invariant measure.

Let us also mention that a derivation of the SQE for the φ_2^4-model on $\mathbb{R}^2$ from a Kac-Ising model has been achieved in [**MW17a**]. This goes back to work initiated in [**FR96**] for $d = 1$ and conjectured results in [**GLP99**] for $d = 2, 3$ (see also [**HI18**]). The SQEs for the exponential/trigonometric models on $\mathbb{R}^2$ have been discussed by Dirichlet forms in [**AKMR23**] and by semigroup methods, similarly as in [**AK20**] (the latter for the φ_3^4-model on $\mathbb{T}^3$). Strong solutions for the trigonometric model on $\mathbb{R}^2$ have been discussed in [**AKMR23**], [**AHR01**], and [**AHR96**] (using Colombeau distributions method, showing in particular the necessity of renormalization) and in [**HS16**] by the theory of regularity structures. The SQE for the exponential model on $\mathbb{R}^2$ has been discussed in [**HKK21**] and [**HKK20**]. Let us also mention that a lot of work has been done in the mass-zero exponential model (called Liouville model), see references in [**AKMR23**], relating to methods of [**Kus92**] and [**Kah85**].

In recent years also other methods to perform stochastic quantization for all known scalar field models for $d = 1$ and $d = 2$ have been developed. One is the method of dimensional reduction from corresponding supersymmetric models. In the case of polynomial models this is discussed in [**KFP83, KLP84**] and in [**ADVG20, ADVG21**]. The latter also permits to cover polynomial and exponential models on $\mathbb{T}^d$ ($d \le 2$) from supersymmetric models in dimension $d+2$, and also an elliptic equation analogue on $\mathbb{T}^d$ ($d \le 2$). For the exponential model optimal range of parameters has been achieved in [**ADVG21**].

Yet another method of stochastic quantization is by the use of nonlocal Dirichlet forms, this has been applied to polynomial models for $d \le 2$ and to the φ_3^4-model (see [**AKYY21**]). For further references concerning the SQE and SQE$_\varepsilon$ for $d = 2$, see [**AMR15**] and [**MR99**]. In a very recent paper [**DPD20**] Da Prato and Debussche have proven estimates on the gradient of the transition semigroup P_t for the SQE to the φ_2^4-model on $\mathbb{T}^2$ and shown that cylinder functions form a core for the generator of this semigroup, in fact the P_t maps bounded Borel functions into Lipschitz continuous functions.

We shall now pass to describe the situation with the study of the SQE for the φ_3^4-model. The problem of giving a rigorous meaning to solutions of this singular SPDE for this model remained open for quite a long time. In [**ALZ06**] it was pointed out that the method of Dirichlet forms as minimally defined as not-yet closed quadratic forms on cylindrical functions would have difficulties in its being carried thorough (because the divergences would make it difficult to show closability in spaces of interest). A breakthrough was realized in work by Martin Hairer on regularity structures [**Hai14**] and in work based on Gubinelli's extension of the method of paracontrolled distributions for singular SPDEs (see [**GIP15**], [**CC18**], [**FG19**] and [**GH19**]). Hairer's method is an innovative use of PDE methods on C^{α}-spaces with negative index. Gubinelli's methods constitute an extensions of T. Lyon's rough paths methods to the case of multidimensional time (see [**Gub04**] and [**Gub10**]). Let us note in passing that since Hairer's work the SPDE for the φ_3^4-model is also called, besides SQE, the equation for the "dynamical φ_3^4-model". The methods of Hairer and Gubinelli apply to many other SPDEs besides the one for the φ_3^4-model, we shall however concentrate ourselves in mentioning papers directly related to the φ_3^4-model (for other applications see, e.g. [**BH20**] and [**FHSX21**]). In the original work of Hairer solutions for the SQE to the φ_3^4-model on $\mathbb{T}^3$ were found on any space C^{α} with $\alpha \in (-2/3, -1/2)$ and with initial conditions in the same space. Various approximations results for the solutions have been found, from other interaction terms [**HX18**] or from a lattice approximation [**HM18a**]. Other proofs of local in time well-posedness of the SQE on $\mathbb{T}^3$ have been obtained, e.g. by renormalization group methods [**Kup16**]. Existence and uniqueness local in time of solutions of this model on $\mathbb{T}^3$ have also been obtained in [**CC18**] by the method of paracontrolled distributions. The extension to global solutions on $\mathbb{T}^3$ was suggested in [**Hai14**] and [**HM18a**] in view of the Bourgain method, and carried through in [**MW17b**]. We shall discuss this reference below, together with other papers that were done roughly at the same time when our paper [**AK20**] was announced in the arXiv (2017). Hairer and Mattingly proved in [**HM18b**] the strong Feller property of the SQE for $\mathbb{T}^3$, for initial data of suitable regularity. In our previous paper [**AK20**] we constructed global solutions of SQE on $\mathbb{T}^3$ using semigroup methods, providing also a proof of invariance of the equilibrium measure (in this sense a new construction of a probability measure associated with the φ_3^4-model on $\mathbb{T}^3$, the measure having the same form as the φ_3^4-model Euclidean measure when reduced to the torus $\mathbb{T}^3$, in fact coinciding with it when a finite-dimensional approximation is considered). In [**AK20**] the solution to the SQE on $\mathbb{T}^3$ was shown to exist in suitable Besov spaces of negative index. The limit process has continuous paths and has a translation invariant φ_3^4-measure as an invariant measure. A local in time uniqueness result is expected to hold on the basis of results in [**Hai14**] and [**CC18**].

In [**MW17b**] Mourrat and Weber proved existence and uniqueness of global solutions in time of the SQEs on the 3-dimensional torus $\mathbb{T}^3 = (\mathbb{R}/2\pi\mathbb{Z})^3$. They also proved the stronger property of "coming down from infinity" of the solutions, in the sense that after a finite time the solutions are contained in a compact set in the state space uniformly with respect to the initial condition. In [**GH19**] Gubinelli and Hofmanová proved the existence and uniqueness of solutions global in the Euclidean space $\mathbb{R}^3$ and in computer time τ of the SQEs, also with an associated "coming down from infinity" property. Their results use the study of the elliptic version of the SQEs on $\mathbb{R}^d$ with $d = 4, 5$ (that have the same kind of singularities as the stochastic quantization equations for $d = 2, 3$). The usefulness of the study of such elliptic equations is inspired by the heuristic dimensional reduction suggested by work of Parisi and Sourlas [**PS79**], partially validated by [**KFP83, KLP84**] in a regularized version of the models. We already recalled above in our discussion of the SQE for the φ_2^4-model that dimensional reduction has been worked out mathematically without regularization for the elliptic case $d = 4$ (corresponding to $d = 2$ in the stochastic quantization equation case) in [**ADVG20**] (see also [**ADVG21**] and [**DVG21**] for this reduction technique, based on the use of supersymmetry). For handling the global problem in $\mathbb{R}^3$ Gubinelli and Hofmanová [**GH19**] introduced Besov spaces with suitable weights. In order to obtain the necessary estimates to gain control on the behavior at spatial infinity of solutions of the equations, they introduced a new localization theory by which they split distributions belonging to weighted spaces into an irregular component behaving well at spatial infinity and a smooth component growing in space. In this way the singular SPDEs under study split into two equations, one containing the irregular component in a linear (or almost linear) way, easy to handle in weighted spaces, and another equation containing all regular terms and all the nonlinear terms that can be handled by PDE arguments adapted to the weighted setting. In the (parabolic) SQE case the uniqueness of solutions is proven essentially by L^2-energy estimates, the property of "coming down from infinity" (proven first in [**MW17b**]) is established by choosing suitable time dependent weights and fully exploiting energy-time estimates (in L^2-spaces). The paper also compares the methods and results obtained with the ones presented in [**MW17b**] (concerning the model on the torus $\mathbb{T}^3$). The global uniqueness problem for the elliptic case remained open.

A subsequent paper [**GH21**] by Gubinelli and Hofmanová presents an alternative construction of a translation invariant φ^4-measure (as invariant measure of the SQE) on $\mathbb{R}^3$ by using methods of the theory of singular SPDEs. Since the results of [**GH21**] are closely related to those in our present paper, we shall describe them in some details in order also to compare them with our methods and results. Gubinelli and Hofmanová construct an invariant measure ν_λ of the SQE for all values of the coupling constant λ as limit points of the tight family $(\nu_{M,\varepsilon})$ (for any $\lambda > 0$ and $m_0 > 0$) and prove that it satisfies the properties of invariance under translations in the underlying Euclidean space $\mathbb{R}^3$, the important axiom of reflection positivity as well as the regularity axiom of OS (in the form of [**OS75**]). These properties are sufficient for the analytic continuation of the moment functions to ν_λ to obtain the corresponding functions in Minkowski space M^3 satisfying the axioms of invariance under the translation group in M^3, positivity (i.e. Hilbert space structure) and regularity (in the sense of corresponding Wightman functions axioms). They do not however prove the rotation invariance property for ν_λ (and hence they cannot deduce Lorentz invariance of the Wightman functions). In addition Gubinelli and Hofmanová prove that their measure ν_λ is non-Gaussian. They also deduce an integration by parts formula and from this they show that the Dyson-Schwinger equations (derived in [**FRa77**], on the basis on previous work on the Euclidean approach to constructive quantum field theory) for the Schwinger functions (moment functions of ν_λ) are satisfied. This construction is very important, since it is streamlined and avoids techniques like cluster expansion, correlation or skeleton inequalities of the constructive quantum field approach. The main advantage (as also of our construction in our previous paper [**AK20**] on the torus model) is to have reflection positivity assured by starting with measures that are finite-dimensional approximations of the one associated with the SQE of the ϕ_3^4-model on a bounded lattice (that is both translation invariant and reflection positive).

The choice of approximation chosen in [**GH21**] is namely of the form

$$\nu_{\lambda,M,\varepsilon} = \nu_{M,\varepsilon}$$
$$= Z_{M,\varepsilon}^{-1} \exp\left\{ -2\varepsilon^d \sum_{\Lambda_{M,\varepsilon}} \left(\frac{\lambda}{4}|\varphi|^4 + \frac{-3\lambda a_{M,\varepsilon} + 3b^2 b_{M,\varepsilon} + m_0^2}{2}|\varphi|^2 + \frac{1}{2}|\nabla_\varepsilon \varphi|^2 \right) \right\}$$
$$\times \prod_{x\in\Lambda_{M,\varepsilon}} d\varphi(x),$$

where ∇_ε denotes the discrete gradient on the periodic lattice $\Lambda_{M,\varepsilon} = (\varepsilon\mathbb{Z}/M\mathbb{Z})^3$, M is an infrared cut-off and ε is an ultraviolet cut-off. $Z_{M,\varepsilon}$ is a normalization constant. $\nu_{M,\varepsilon}$ is both translation and reflection-positive (the latter can be seen by methods discussed e.g. in [**GJ81**, Theorem 7.10.3] and [**Sim74**, Theorem VIII.15.]). The removal $\varepsilon \downarrow 0$ and $M \to \infty$ is obtained by choosing appropriately the counterterms $a_{M,\varepsilon}$ and $b_{M,\varepsilon}$. The essential point is the proof of the tightness of the family $(\nu_{M,\varepsilon})_{M,\varepsilon}$ as embedded in $\mathcal{S}'(\mathbb{R}^3)$. This is achieved by looking at the analogue on the periodic lattice $\Lambda_{M,\varepsilon}$ of the SQE for the ϕ_3^4-model. This analogue constitutes a finite-dimensional system of the SDE which is globally well-posed and has $\nu_{M,\varepsilon}$ as its unique invariant measure. In this sense this corresponds to our method of studying the SQE on $\mathbb{T}^3$ in our previous paper [**AK20**]. But in the case of [**GH21**] an extension of a renormalized energy method (similar to the one in [**AK20**] for the case of $\mathbb{T}^3$) is used to cope with the case of the weighted Sobolev spaces needed to handle the global problem in $\mathbb{R}^3$. This extension of the renormalized energy method is based on the previous other work by Gubinelli and Hofmanová (we already mentioned above [**GH19**]), in turn using work in [**MP19**] on Besov spaces on the lattices. As compared to [**GH19**] in the paper [**GH21**] there is a more detailed control on the solutions of the SQE, both stationary and non-stationary. This leads in particular to establish that any accumulation point ν_λ of the tight family $(\nu_{M,\varepsilon})_{M,\varepsilon}$ (for any $\lambda > 0$ and $m_0^2 > 0$) is translation invariant, reflection-positive and non-Gaussian. In addition, for any small $\kappa > 0$, Gubinelli and Hofmanová prove that there exists $\sigma > 0$, $\beta > 0$ and $v = O(\kappa) > 0$ such that the stretched exponential integrability

$$\int_{\mathcal{S}'(\mathbb{R}^3)} \exp\left\{\beta \|(1+|\cdot|^2)^{-\sigma}\varphi\|^{1-v}_{H^{-1/2-\kappa}}\right\} \nu(d\varphi) < \infty$$

holds. The authors of [**GH21**] show that this is enough to establish the regularity axiom of [**OS75**] for the Schwinger functions in the form

$$|S_n(f_1, f_2, \dots, f_n)| \le k^n (n!)^\beta \prod_{i=1}^n \|f_i\|_{H^{1/2+2\kappa}(\rho^{-2})}.$$

Moreover, they show that ν_λ satisfies an integration by parts formula leading to the hierarchy of Dyson-Schwinger equations for the Schwinger functions already mentioned above. Let us also note that by this construction the authors provide an interpretation of the cubic term in the integration by parts as being given by a random distributions (the problem of giving a meaning to that term was mentioned before in [**ALZ06**]). All results hold for all $\lambda > 0$ and $m_0^2 > 0$. Moreover an extension of the construction for a fractional variant of the model is provided, with $-\triangle$ replaced by $(-\triangle)^\gamma$ ($\gamma \in (21/22, 1)$), and their argument is naturally extendable to multicomponent models with $O(N)$-symmetry. See [**SZZ22**] for connections with the $O(N)$ linear sigma model for large N (on the torus $\mathbb{T}^3$).

Let us also mention a few other papers that also appeared after our submission of [**AK20**]. In [**MW20**] Mourrat and Weber established results similar to those in [**GH21**, Section 4]. The authors proved a priori bounds for the solutions of the ϕ_3^4-SQE. Control on a compact space-time set was obtained independently of boundary conditions. They pointed out that their bounds could be used in a compactness argument to construct solutions on the full space $\mathbb{R}^3$ and invariant measures.

In [**BG20**] a variational approach for the construction of the finite volume ϕ_3^4-measure ν has been presented and shown to hold. The method used in that paper is based on methods of stochastic control: in fact that the Laplace transform of an approximation by regularization of the φ_3^4-measure ν_T (with renormalization terms) on the 3-torus is interpreted as the value function of a stochastic optimal control problem along the flow of the scale regularization parameter T. Under removal of the regularization parameter (i.e. $T \to \infty$) it is shown by Γ-convergence (of variational functionals) that a limit probability measure ν exists whose Laplace transform is also characterized by a variational principle. This is an important result, providing an explicit formula also for ν as φ_3^4-measure on the 3-torus. Let us remark that this approach was also used in successive work in [**ORSW20**] for a study of the SQE for the hyperbolic Sine-Gordon model on $\mathbb{T}^2$ (for this model, in the Euclidean version, see also the recent work [**AKMR23**] and references therein).

In [**BG21**] Barashkov and Gubinelli show that the ϕ_3^4-measure on $\mathbb{T}^3$ is absolutely continuous with respect to a new measure constructed by a random shift of Nelson's Gaussian free field. To show this the authors prove a Girsanov type theorem with respect to the filtration generated by a scale parameter. As a byproduct they prove the long standing conjecture (see e.g. [**Fel74**] and [**ALZ06**]) that the ϕ_3^4-measure (on $\mathbb{T}^3$) is singular with respect to Nelson's Gaussian free field measure. For achieving this a Brownian motion $(W_t)_{t\geq 0}$ with values in $\mathcal{S}'(\mathbb{T}^3)$ is considered, as a regularization of Nelson's Gaussian free field measure μ on $\mathbb{T}^3$ "at scale t". The authors also show that the ϕ_3^4-measure ν on $\mathcal{S}'(\mathbb{T}^3)$ is the weak limit for $T \to \infty$ of the path space measure $\mathbb{P}^T$ of $(W_t)_{t\geq 0}$, with density $Z_T^{-1}\exp(-V_T(W_T))$ with respect to μ and $V_T(\varphi) := \lambda \int (\varphi(x)^4 - a_T\varphi(x) + b_T)dx$, where (a_T, b_T) is a family of appropriate diverging (for $T \to +\infty$) counterterms. An adapted process v with values in $L^2(\mathbb{T}^3)$ is constructed and a Girsanov transformed measure $\mathbb{Q}^v$ such that W_t is a solution of the path dependent SDE with a polynomial drift term of the form $V_t(\tilde{W}_s, s \in [0,t])$, with $\tilde{W}_t$ a Gaussian martingale under $\mathbb{Q}^v$. $\mathbb{Q}^v$ can then be looked upon as a natural reference measure for the ϕ_3^4-measure on $\mathbb{T}^3$, instead of the free field measure μ, that is shown to be singular both for $\mathbb{Q}^v$ and for ν. This construction is expected to have interesting applications also to many other areas.

Let us now describe the structure of the present paper. In Section 1.2 we present the setting of our work and the main theorems (Theorems 1.1 and 1.2) followed by a comparison of our work with the results in [**GH21**] and [**AK20**]. The setting of the present work is exposed at the beginning of Section 1.2 starting by explaining the notation for our finite-dimensional version $\mu_{M,N}$ (see (1.9) and (1.10) below) of the ϕ_3^4-measure μ that we intend to construct. This contains, in the terminology discussed before in this introduction, both an "infrared cut-off" expressed by the smooth non-negative function $\rho_M(x)$ that cut-off the integral in the space variable $x \in \mathbb{R}^3$ in (1.9) at values $|x| \geq 2M$ ($M \in \mathbb{N}$) and an "ultraviolet cut-off" expressed by the presence in (1.9) of $P_{M,N}$ given by $\rho_M(\cdot)P_N$, with P_N such that the Fourier transform of its action on a Borel function $f \in L^2(\mathbb{R}^3;\mathbb{C})$ amounts to the multiplication by $\psi_N(\xi)$, with ψ_N an even, positive, smooth function on $\mathbb{R}^3$ that vanishes for values in the Fourier transform variable ξ such that $|\xi| \geq 2^{N+1}$. The measure $\mu_{M,N}$ is thus of Gibbs-type, with a doubly regularized Gibbsian function $U_{M,N}$, and reference measure μ_0, namely Nelson's free field measure with a positive mass m_0, i.e. the centered Gaussian measure on $\mathcal{S}'(\mathbb{R}^3)$ with the covariance operator $[2(-\triangle + m_0^2)]^{-1}$. $U_{M,N}$ contains the coupling constant $\lambda \in (0, \lambda_0]$ with $\lambda_0 > 0$ given, and renormalization terms $C_1^{(N)}$ and $C_2^{(M,N)}$ chosen to be divergent as $M, N \to \infty$ in such a way as to compensate the divergence of the fourth power term, making possible that $\mu_{M,N}$ converges to the candidate μ for the ϕ_3^4-measure. The difference between our approximation of μ as compared to the one used by [**GH21**] is that in our case $C_2^{(M,N)}$ depends on $x \in \mathbb{R}^3$, whereas in the translation invariant approximation used in [**GH21**] the analogue of $C_2^{(M,N)}$ is independent of x. As we will see our choice has the advantage of assuring rotation invariance for the limit point μ.

Our main results are Theorems 1.1 and 1.2. Theorem 1.1 asserts that for any $\lambda > 0$ and $m_0^2 > 9/2$ (this can be relaxed to $m_0^2 > 0$, see Remark 1.3), there is a subsequence $M(k)$ and $N(k)$ for $k \in \mathbb{N}$ of $M, N \to \infty$ such that $\mu_{M(k),N(k)}$ converges weakly to a limit probability measure μ. Moreover, μ is a stationary measure of a continuous process X on a certain intersection of weighted Besov spaces with negative regularity index $-(1/2) - \varepsilon$ ($\varepsilon > 0$) and with weight $\nu(x) = (1 + |x|^2)^{-\sigma/2}$ ($\sigma \in (3, \infty)$). X is obtained as a limit of the processes $X_t^{M,N}$ given by (1.11) below, with initial law given by $\mu_{M,N}$, as $M, N \to \infty$. Theorem 1.2 states that if the cut-off functions ρ and ψ are chosen to be rotation invariant, then μ is invariant with respect to rotations and reflections in the underling space $\mathbb{R}^3$. It is also shown that the support of μ is contained in the weighted Besov space $B_p^{-(1/2)-\varepsilon}(\nu^{p/2})$ for all $p \in [2, \infty)$. Moreover, the square moment estimate (iii) in Theorem 1.2 holds. In addition we show that our limit measure μ is not Gaussian on $\mathcal{S}'(\mathbb{R}^3)$ and is reflection positive.

Our method at present does not lead to growth bounds for the moment functions of μ, as were established in [**GH21**, Theorem 1.1] for their construction of ν_λ. The reason is that in the present paper the approximation operators P_N and ρ_M are mixed in the stochastic quantization equation (see (1.11) below) for the approximation measures, while such approximation operators do not appear in the case of approximations by discretization and tori (see [**GH21**, (3.1)]). We remark that this difference comes from the approximations. As in [**GH21**] the cluster property of the limit measure μ is not discussed as well as the question of whether the corresponding relativistic ϕ_3^4 would yield (at least for small values of λ) nontrivial scattering. As we recalled above, these additional properties have been proven for the Euclidean measures constructed via the cluster expansion (see [**FO76**] and [**MS76**]). We expect that our limit measure μ at least for small values of λ will also be translation invariant. The question about identification of our limit measure with the one constructed in [**GH21**] or, more generally, a uniqueness result for reflection positive, rotation and translation invariant probability measures that are invariant measures for the SQE (1.7) and would coincide with the Euclidean measure of [**FO76**] and [**MS76**] (at least in the "weak coupling limit") have to be left for future investigations.

After the preprint version of the present paper appeared, this topic continues to be studied and a lot of new papers have appeared. We mention some of them most connected to the present paper: [**ABDVG22**] is about the stochastic quantization of Euclidean fermionic quantum field theories whereas [**CCHS22**] the stochastic quantization of the Yang-Mills-Higgs model. [**BDV21**] discusses the elliptic stochastic quantization for the sinh-Gordon quantum field theory and [**HS22**] the sub-Gaussian tails for the Φ^4_3-measure on the torus. [**SZZ21**] connects the perturbation theory for the Φ^4_2-model with its stochastic quantization. In particular, the type of approximation in the present paper is applied in [**SZZ21**] for the Φ^4_2-model to be considered on the whole space.

In the next section we shall provide the precise setting and the statement of the main results in the present paper.

1.2. Setting and Main theorems

In this section we give the precise setting and the statement of the main results in the present paper.

Denote the spaces of Schwartz-class functions and of tempered distributions by $\mathcal{S}(\mathbb{R}^3)$ and $\mathcal{S}'(\mathbb{R}^3)$, respectively, and denote this pairing by $\langle \cdot, \cdot \rangle$. We remark that the pairing $\langle \cdot, \cdot \rangle$ of $\mathcal{S}(\mathbb{R}^3)$ and $\mathcal{S}'(\mathbb{R}^3)$ is given by the extension of the inner product on $L^2(dx) := L^2(\mathbb{R}^3, dx)$. Let ψ be a non-negative C^∞-function on $\mathbb{R}^3$ such that $\psi(\xi) = 1$ for $|\xi| \le 1$, $\psi(\xi) = 0$ for $|\xi| \ge 2$ and $\psi(\xi) = \psi(-\xi)$ for $\xi \in \mathbb{R}^3$, and let ρ be a non-negative C^∞-function on $\mathbb{R}^3$ such that $\rho(x) = 1$ for $|x| \le 1$ and $\rho(x) = 0$ for $|x| \ge 2$.

Let $\mathcal{F}$ and $\mathcal{F}^{-1}$ be the Fourier transform and the inverse Fourier transform operators on $\mathbb{R}^3$, respectively (see Chapter 2 for details). For $M, N \in \mathbb{N}$ define P_N, $P_{M,N}$ and $P^*_{M,N}$ as the mappings from $\mathcal{S}'(\mathbb{R}^3)$ to $\mathcal{S}'(\mathbb{R}^3)$ given by

$$\begin{aligned} P_N f(x) &:= \mathcal{F}^{-1}\left[\psi_N(\mathcal{F}f)\right](x), \\ P_{M,N} f(x) &:= \rho_M(x) P_N f(x), \\ P^*_{M,N} f(x) &:= P_N(\rho_M f)(x), \end{aligned}$$

for $f \in \mathcal{S}'(\mathbb{R}^3)$, where

$$\begin{aligned} \psi_N(\xi) &:= \psi(2^{-N}\xi), \quad \xi \in \mathbb{R}^3 \\ \rho_M(x) &:= \rho(M^{-1}x), \quad x \in \mathbb{R}^3. \end{aligned}$$

Note that $P^*_{M,N}$ is the dual operator of $P_{M,N}$ with respect to $L^2(dx)$.

Let $\lambda_0 \in (0, \infty)$ and let $\lambda \in (0, \lambda_0]$ be fixed. Let $m_0 \in (0, \infty)$ be a fixed constant interpreted as the mass of the underling Euclidean free field. Define $C_1^{(N)}$ and $C_2^{(M,N)}$ by

$$\begin{aligned} C_1^{(N)} &:= \left\langle [2(m_0^2 - \triangle)]^{-1} P_N^2 \delta_0, \delta_0 \right\rangle \\ C_2^{(M,N)}(x) &:= 2 \sum_{i,j=-1}^{\infty} \mathbb{I}_{[-1,1]}(i-j) \\ &\quad \times \int_0^\infty e^{t(\triangle - m_0^2)} P_N^2 \left[\left(\Delta_i P_N \left[\rho_M^2 V_N(t) \left(\rho_M^2 \Delta_j P_N \delta_x \right) \right] \right)^2 \right] (x) dt \end{aligned}$$

where δ_x is the Dirac delta function with support at x, $\triangle$ is the Laplacian on $\mathbb{R}^3$, $\{\Delta_j ; j \in \mathbb{N} \cup \{-1, 0\}\}$ is the (Littlewood-Paley) nonhomogeneous dyadic blocks (see

Chapter 2 for details), and

$$V_N(t) := P_N^2[2(m_0^2 - \triangle)]^{-1} e^{t(\triangle - m_0^2)}.$$

Define a function $U_{M,N}$ on $\mathcal{S}'(\mathbb{R}^3)$ by

(1.9)
$$U_{M,N}(\phi) := \int_{\mathbb{R}^3} \left\{ \frac{\lambda}{4}(P_{M,N}\phi)(x)^4 - \frac{3\lambda}{2}\left(C_1^{(N)} - 3\lambda C_2^{(M,N)}(x)\right)\rho_M(x)^2(P_{M,N}\phi)(x)^2 \right\} dx,$$

for $\phi \in \mathcal{S}'(\mathbb{R}^3)$, and consider the probability measure $\mu_{M,N}$ on $\mathcal{S}'(\mathbb{R}^3)$ given by

(1.10)
$$\mu_{M,N}(d\phi) = Z_{M,N}^{-1} \exp\left(-U_{M,N}(\phi)\right)\mu_0(d\phi),$$

where $Z_{M,N}$ is the normalizing constant (so that the total mass of $\mu_{M,N}$ is 1), and μ_0 is the free field measure with a positive mass m_0, i.e. the centered Gaussian measure on $\mathcal{S}'(\mathbb{R}^3)$ with the covariance operator $[2(-\triangle + m_0^2)]^{-1}$. We remark that $\{\mu_{M,N}\}$ is an approximation sequence for the Φ_3^4-measure, which will be constructed below as an invariant probability measure of the flow associated with the stochastic quantization equation, and $C_1^{(N)}$ and $C_2^{(M,N)}(x)$ are provided for renormalization. We also remark that $C_1^{(N)}$ is independent of M and $x \in \mathbb{R}^3$, while $C_2^{(M,N)}(x)$ depends on M and $x \in \mathbb{R}^3$. The dependence of the second renormalization term $C_2^{(M,N)}(x)$ on M and x comes from our approximation in space by ρ_M. Let us remark that for a torus approximation of $\mathbb{R}^3$ as in [**GH21**] the analogue of $C_2^{(M,N)}$ can be taken to be independent of x, because of translation invariance. This is however not the case with the choice of our approximation by ρ_M.

Consider the stochastic partial differential equation (SPDE)

(1.11)
$$\begin{cases} \partial_t X_t^{M,N}(x) &= \dot{W}_t(x) - (-\triangle + m_0^2)X_t^{M,N}(x) \\ &\quad -\lambda P_{M,N}^* \Big\{ (P_{M,N}X_t^{M,N})^3(x) \\ &\qquad -3\left(C_1^{(N)} - 3\lambda C_2^{(M,N)}(x)\right)\rho_M(x)^2 P_{M,N}X_t^{M,N}(x) \Big\}, \\ X_0^{M,N}(x) &= \xi_{M,N}(x), \quad t \in [0,\infty), \ x \in \mathbb{R}^3 \end{cases}$$

where the initial condition $\xi_{M,N}$ is a $\mathcal{S}'(\mathbb{R}^3)$-valued random variable with law $\mu_{M,N}$ and $\dot{W}_t(x)$ is a (centered) Gaussian white noise (in the variable t and in the space variable $x \in \mathbb{R}^3$) independent of $\xi_{M,N}$. The equation (1.11) comes from the stochastic quantization equation corresponding to the measure $\mu_{M,N}$ with initial condition $\xi_{M,N}$. From now on we shall assume that

(1.12)
$$m_0^2 > \frac{9}{2}.$$

This assumption can be relaxed to $m_0^2 > 0$ by modifying ν defined below. For detail, see Remark 1.3 below. Here and below, we shall also use weighted Besov spaces. Let ν be a smooth function on $\mathbb{R}^3$ such that ν is positive everywhere, and denote the L^p-space and Besov spaces with weight function ν by $L^p(\nu)$ and $B_{p,r}^s(\nu)$, respectively (see Chapter 2 for the details of the definition of weighted Besov spaces).

Let us take $\nu(x) := (1+|x|^2)^{-\sigma/2}$ for $x \in \mathbb{R}^3$ and some $\sigma \in (3,\infty)$. Then, $\nu \in L^1(dx)$ and the unique solution $X_t^{M,N}$ to (1.11) is a $B_2^{-\alpha}(\nu)$-valued stationary Markov process for $\alpha \in (1/2,\infty)$ (see Theorem 4.1).

In the setting above, in particular the above choice of ν, we have the following main results.

THEOREM 1.1. *Let $\{M_N; N \in \mathbb{N}\}$ be a given $\mathbb{N}$-valued sequence such that $\lim_{N\to\infty} M_N = \infty$ and*

$$\lim_{N\to\infty} M_N 2^{-\delta N} = 0$$

for all $\delta \in (0,1]$. Then, for $\varepsilon \in (0,1/16]$, there exists a sequence $\{X^{M_{N(k)},N(k)}; k \in \mathbb{N}\}$ such that $\lim_{k\to\infty} N(k) = \infty$ and $\{X^{M_{N(k)},N(k)}; k \in \mathbb{N}\}$ converges in law on $C([0,\infty); B_{4/3}^{-1/2-\varepsilon}(\nu) \cap B_{12/5}^{-1/2-\varepsilon}(\nu^{9/5}))$. Moreover, if X is the limit in law of $\{X^{M_{N(k)},N(k)}; k \in \mathbb{N}\}$ on $C([0,\infty); B_{4/3}^{-1/2-\varepsilon}(\nu) \cap B_{12/5}^{-1/2-\varepsilon}(\nu^{9/5}))$, then X is a continuous stationary process on $B_{4/3}^{-1/2-\varepsilon}(\nu) \cap B_{12/5}^{-1/2-\varepsilon}(\nu^{9/5})$ and the limit probability measure μ in the weak convergence sense of the associated subsequence $\{\mu_{M_{N(k)},N(k)}\}$ is a stationary measure of X.

The assumption on M_N is provided for M_N to diverge sufficiently slowly compared with N. Since $\{\mu_{M,N}\}$ is a proper two-parameter approximation sequence of the Φ_3^4-measure on $\mathbb{R}^3$, the measure μ constructed in Theorem 1.1 is our construction of the Φ_3^4-measure (that we shall call henceforth for simplicity the Φ_3^4-measure). The stochastic process X constructed in Theorem 1.1 is a flow associated to μ. Moreover, we are able to show that the Φ_3^4-measure μ constructed in Theorem 1.1 has the following properties.

THEOREM 1.2. *Let μ be the measure obtained in Theorem 1.1. Then, μ has the following properties:*

(i) *If the functions ψ and ρ, which are used to define the approximation of the Φ_3^4-measure, are chosen to be radially symmetric, then μ is rotation and reflection invariant, i.e.*

$$\int f(\phi)\mu(d\phi) = \int f(\theta\phi)\mu(d\phi), \quad f \in C_b(\mathcal{S}'(\mathbb{R}^3)),\ \theta \in \mathrm{O}(3)$$

where $\mathrm{O}(3)$ is the orthogonal group on $\mathbb{R}^3$ and $(\theta\phi)(x) := \phi(\theta x)$ for $x \in \mathbb{R}^3$.

(ii) *For $f \in \mathcal{S}(\mathbb{R}^3)$, define $f^\dagger \in \mathcal{S}(\mathbb{R}^3)$ by $f^\dagger(x_1,x_2,x_3) := f(-x_1,x_2,x_3)$ for $(x_1,x_2,x_3) \in \mathbb{R}^3$. If $\psi^\dagger = \psi$ and $\rho^\dagger = \rho$, then μ has the reflection positivity property i.e.*

$$\int \overline{F^\dagger(\phi)} F(\phi)\mu(d\phi) \geq 0$$

($\bar{\cdot}$ means complex conjugation) for any $F \in C_b(\mathcal{S}'(\mathbb{R}^3);\mathbb{C})$ given by

$$F(\phi) = \sum_{j=1}^n a_j e^{\sqrt{-1}\langle f_j,\phi\rangle}$$

for some $a_1,a_2,\dots,a_n \in \mathbb{C}$, $f_1,f_2,\dots,f_n \in \mathcal{S}(\mathbb{R}^3_+)$ and $n \in \mathbb{N}$, where $\mathcal{S}(\mathbb{R}^3_+)$ is the set of functions $f \in \mathcal{S}(\mathbb{R}^3)$ with support in $\{(x_1,x_2,x_3); x_1 > 0\}$, and $F^\dagger$ is defined by

$$F^\dagger(\phi) = \sum_{j=1}^n a_j e^{\sqrt{-1}\langle f_j^\dagger,\phi\rangle}.$$

(iii) *For each* $p \in [2, \infty)$

$$\int \|\phi\|^2_{B_p^{-1/2-\varepsilon}(\nu^{p/2})} \mu(d\phi) < \infty.$$

In particular, the support of μ *is included in* $B_p^{-1/2-\varepsilon}(\nu^{p/2})$ *for all* $p \in [2, \infty)$.

(iv) μ *is not a Gaussian measure on* $\mathcal{S}'(\mathbb{R}^3)$. *Here, the definition that a measure* μ *on* $\mathcal{S}'(\mathbb{R}^3)$ *is a Gaussian measure is that for any* $n \in \mathbb{N}$ *and* $f_1, f_2, \dots, f_n \in \mathcal{S}(\mathbb{R}^3)$, $(\langle f_1, \phi\rangle, \langle f_2, \phi\rangle, \dots, \langle f_n, \phi\rangle)$ *has an* n*-dimensional Gaussian distribution under* μ.

Theorem 1.2 implies that if ψ and ρ are radially symmetric, our Φ^4_3-measure constructed in Theorem 1.1 is nontrivial (i.e. non-Gaussian), rotation invariant and reflection positive.

REMARK 1.3. The assumption (1.12) is relaxed to $m_0^2 > 0$ by replacing $\nu(x) = (1+|x|^2)^{-\sigma/2}$ with $\nu_a(x) = (1+a|x|^2)^{-\sigma/2}$ for some $a \in (0, 2m_0^2/9)$. We introduced (1.12) to choose σ so that (4.1) holds, and (4.1) is applied in the proof of Proposition 6.5. However, by the replacement ν by ν_a for such a, we are able to choose σ so that

$$9 < \sigma^2 < \frac{2m_0^2}{a^2}$$

instead of (4.1). This enable us to obtain (6.10) with some positive constant c, because

$$|\nabla \log \nu_a(x)| = \frac{\sigma a|x|}{1+a|x|^2} \leq \frac{\sigma\sqrt{a}}{2}.$$

We got this technique from Professors Rongchan Zhu and Xiangchan Zhu.

As compared with the results in [**GH21**] we have the same statement of non-Gaussianity of the limit measure μ and its reflection positivity, but in addition we have also the rotation (and reflection) invariance of the limit measure, however we do not discuss here translation invariance (a property assured instead in [**GH21**] for their limit measure). The main difference between our method of approximation and the discrete approximation by Gubinelli and Hofmanová that uses a doubly scaled torus, consists in the fact that we avoid the use of a torus but have instead an approximation by a localization of the interaction (by having a space cut-off). We would have to regain the translation invariance in the limit, but we keep the rotation invariance all the way along, as opposite to Gubinelli and Hofmanová. Also the interplay of cut-off and weights are reflected by technical differences in the construction of the relevant Besov spaces in our construction and in the one of Gubinelli and Hofmanová [**GH21**]. Moreover, also the use of the estimates on the approximations of the Ornstein-Uhlenbeck processes and their powers that we need are similar but different from the torus estimates of [**GH21**]. Here we also use a technique similar to the one we developed in [**AK20**, Theorem 4.1] to transform (1.11) to another partial differential equation with random coefficients which can be taken to a limit as $M, N \to \infty$, by introducing a pair of random variables $(\xi_{M,N}, \zeta)$ with law invariant for the system $(Y_t^{M,N}, Z_t)$, where Z_t is the Ornstein-Uhlenbeck process discussed in Chapter 3. We consider the SPDE (1.11) for a process $X_t^{M,N}$ with initial condition $\xi_{M,N}$ and a Gaussian white noise independent of $(\xi_{M,N}, \zeta)$ (the same as the one used for the definition of Z_t). We then consider $X_t^{M,N} - Z_t$ and write an SPDE for it, (4.7) below, containing the terms $\mathcal{Z}_t^{(k,N)}$ discussed in Chapter 3. And we iterate the procedure, so to get estimates suitable for applying the theory of paraproducts (as first applied to singular SPDEs in work by Gubinelli and coworkers). We get part of the necessary estimates following our previous work [**AK20**], but we meet the problem that our approximation, by the use of P_N, is not symmetric on $L^2(\nu)$. We overcome this problem by controlling the relevant commutators between the weight functions for the Besov spaces and the approximations we use.

The organization of the other sections of the present paper is as follows. Chapter 2 is provided in order to prepare the function spaces and the inequalities which we apply in the proofs of the main theorems. In Section 2.1, we survey weighted Besov spaces and paraproducts. In Section 2.2, we present some properties of our approximation operators and commutator estimates. These properties imply that our approximation is sufficiently good for applying methods for singular stochastic partial differential equations. We remark that similar properties are also discussed in [**GH21**] within a different setting.

In Chapter 3 we provide the convergence and estimates for some polynomials of the infinite-dimensional Ornstein-Uhlenbeck processes, which appear in the transformation in Chapter 4 of the stochastic quantization equations with respect to the approximation measures of the Φ_3^4-measure. In the present paper, we are concerned with approximations not on a torus, but on the whole space $\mathbb{R}^3$. So we discuss the estimates of the polynomials of the Ornstein-Uhlenbeck processes on weighted Besov spaces by using the heat semigroup and corresponding Green kernel in the underlying space $\mathbb{R}^3$, while a similar argument was done in [**AK20**] by the Fourier expansion on the torus $\mathbb{T}^3$.

In Chapter 4, similarly as in our previous paper [**AK20**], we consider stochastic quantization equations with respect to the approximation measures of the Φ_3^4-measure. The handling of these equations is based on methods related to the theory of the regularity structure and the paracontrolled calculus, and techniques to handle the singular stochastic partial differential equations with locally Lipschitz coefficients.

In Chapter 5 we prepare some estimates for the proofs in Chapter 6 on the bounds of the solutions to the approximation equations for the limit-behaviour in the parameter for the smoothing. The estimates obtained in this section imply that under the localization of interaction, the solutions to the approximation equations are asymptotically the same in the parameter for the smoothing as for the Ornstein-Uhlenbeck process. The estimates in this section did not appear in [**AK20**]; we deduce and apply them to control the commutators between the weight functions of weighted Besov spaces and the approximation operators in the next section.

In Chapter 6 we give a uniform estimate for some functionals of the solutions to the approximation equations and finally prove Theorem 1.1. Many parts of the proof are similar to those in [**AK20**]. However, as mentioned above, unlike [**AK20**] the commutators between the weight functions of weighted Besov spaces and the approximation operator appear. In Chapter 6 we omit the proofs of the estimates which are same as those in [**AK20**], and only concentrate ourselves on the estimates that are different from those in [**AK20**], because of the differences in the frameworks.

In Chapter 7 we prove the properties in Theorem 1.2 of our Φ_3^4-measure. As proved in Section 7.1, we have the rotation and reflection invariance as an advantage of our approximation of the Φ_3^4-measure. In Sections 7.2, 7.3 and 7.4 we prove the reflection positivity, a result on the support of our Φ_3^4-measure and prove the non-Gaussian character of the constructed measure, respectively.

Acknowledgments

The authors are very grateful to Professor Massimiliano Gubinelli and Francesco De Vecchi for stimulating discussions on topics related to this work, to Professors Rongchan Zhu and Xiangchan Zhu for giving us the technique in Remark 1.3, to Hirotatsu Nagoji for finding mistakes in early versions, and Paweł Duch and Professor Ismael Bailleul for sending comments on our argument on reflection positivity.

The first named author gratefully acknowledges support by CIB (where our collaboration on [**AK20**] and the present paper was initiated).

The second named author would like to thank the Isaac Newton Institute for Mathematical Science for an invitation and hospitality during the programme "Scaling limit, rough paths, quantum field theory". He is also grateful to the Institute of Applied Mathematics, and the Hausdorff Center of Mathematics of the University of Bonn for hospitality. The second author is also grateful to Professors David Brydges, Martina Hofmanová, Masato Hoshino, Yuzuru Inahama, Hiroshi Kawabi, Yoshio Tsutsumi and Baris Ugurcan for helpful discussions and valuable comments.

The authors also thank the anonymous referee for one's careful reading and useful comments.

CHAPTER 2

Weighted Besov spaces, paraproducts and estimates of functions

2.1. Weighted Besov spaces and paraproducts

Let dx be the 3-dimensional Lebesgue measure on $\mathbb{R}^3$. Let $L^p(\nu)$ be the pth-order integrable function space with respect to the measure $\nu(x)dx$, where ν is a positive Borel function on $\mathbb{R}^3$ and $p \in [1, \infty]$. We remark that in the present paper the $L^\infty(\nu)$-norm is defined by

$$\|f\|_{L^\infty(\nu)} := \text{ess.sup}_{x\in\mathbb{R}^3}|f(x)| \quad \text{with respect to } dx.$$

Let χ and φ be functions in $C^\infty([0,\infty);[0,1])$ such that the supports of χ and φ are included by $[0,4/3)$ and $[3/4,8/3]$ respectively, and that

$$\chi(r) + \sum_{j=0}^{\infty} \varphi(2^{-j}r) = 1, \quad r \in [0,\infty).$$

We remark that χ and φ satisfy that

$$\varphi(2^{-j}r)\varphi(2^{-k}r) = 0, \quad r \in [0,\infty), \ j,k \in \mathbb{N}\cup\{0\} \text{ such that } |j-k| \geq 2,$$
$$\chi(r)\varphi(2^{-j}r) = 0, \quad r \in [0,\infty), \ j \in \mathbb{N}.$$

For the existence of χ and φ, see [**BCD11**, Proposition 2.10]. Throughout this paper, we fix χ and φ, and we do not mention explicitly this dependence.

Let $\mathcal{S}(\mathbb{R}^3)$ and $\mathcal{S}'(\mathbb{R}^3)$ be the Schwartz (test function) space and the space of tempered distributions on $\mathbb{R}^3$, respectively. We define the (Littlewood-Paley) nonhomogeneous dyadic blocks $\{\Delta_j; j \in \mathbb{N}\cup\{-1,0\}\}$ by

$$\Delta_{-1}f(x) = \left[\mathcal{F}^{-1}\left(\chi(|\cdot|)\mathcal{F}f\right)\right](x), \qquad x \in \mathbb{R}^3$$
$$\Delta_j f(x) = \left[\mathcal{F}^{-1}\left(\varphi(2^{-j}|\cdot|)\mathcal{F}f\right)\right](x), \quad x \in \mathbb{R}^3, \ j \in \mathbb{N}\cup\{0\},$$

where $\mathcal{F}$ and $\mathcal{F}^{-1}$ are the Fourier transform and inverse Fourier transform operators respectively, i,e. $\mathcal{F}$ is the automorphism of $\mathcal{S}'(\mathbb{R}^3)$ given by the extension of the map

$$g \mapsto \widehat{g}(\xi) = \frac{1}{(2\pi)^{3/2}} \int_{\mathbb{R}^3} g(x) e^{-\sqrt{-1}x\cdot\xi} dx, \quad g \in \mathcal{S}(\mathbb{R}^3),$$

where $x\cdot\xi := \sum_{j=1}^3 x_j\xi_j$ for $x = (x_1,x_2,x_3), \xi = (\xi_1,\xi_2,\xi_3) \in \mathbb{R}^3$, and $\mathcal{F}^{-1}$ is the inverse operator of $\mathcal{F}$, respectively (see [**BCD11**, Section 1.2]). As a family of pseudo-differential operators, $\{\Delta_j; j \in \mathbb{N}\cup\{-1,0\}\}$ is given by

$$\Delta_{-1}f = \chi\left(\sqrt{-\triangle}\right)f, \quad \Delta_j f = \varphi\left(2^{-j}\sqrt{-\triangle}\right)f \quad j \in \mathbb{N}\cup\{0\}$$

where $\triangle$ is the Laplace operator on $\mathbb{R}^3$.

We define the weighted Besov norm $\|\cdot\|_{B^s_{p,r}(\nu)}$ and the weighted Besov space $B^s_{p,r}(\nu)$ on $\mathbb{R}^3$ with $s \in \mathbb{R}$ and $p, r \in [1, \infty]$ by

$$\|f\|_{B^s_{p,r}(\nu)} := \begin{cases} \left(\sum_{j=-1}^{\infty} 2^{jsr} \|\Delta_j f\|^r_{L^p(\nu)} \right)^{1/r}, & r \in [1, \infty), \\ \sup_{j \in \mathbb{N} \cup \{-1, 0\}} 2^{js} \|\Delta_j f\|_{L^p(\nu)}, & r = \infty, \end{cases}$$

$$B^s_{p,r}(\nu) := \{f \in \mathcal{S}'(\mathbb{R}^3); \|f\|_{B^s_{p,r}(\nu)} < \infty\}.$$

It is easy to see that if the total mass of the measure $\nu(x)dx$ is finite, we have $B^{s_1}_{p_1,r_1}(\nu) \subset B^{s_2}_{p_2,r_2}(\nu)$ for $s_1, s_2 \in \mathbb{R}$ and $p_1, p_2, r_1, r_2 \in [1, \infty]$ such that $s_1 \geq s_2$, $p_1 \geq p_2$ and $r_1 \leq r_2$. Similarly to the Besov spaces without weights, one has that $B^s_{p,\infty}(\nu) \subset B^{s'}_{p,1}(\nu)$ for $p \in [1, \infty]$ and $s, s' \in \mathbb{R}$ such that $s' < s$ (see [**BCD11**, Corollary 2.96]).

REMARK 2.1. Our definition of weighted Besov spaces is different from those in [**GH19**], but the same as in [**MW17c**].

Below we often assume that $\nu(x) = (1 + |x|^2)^{-\sigma/2}$ for some $\sigma \in [0, \infty)$. It is known that the weights of this type are good for function spaces (see e.g. [**HL18**, Section 3]). We give a relation between Besov spaces and Sobolev spaces as a proposition.

PROPOSITION 2.2. *Assume that $\nu(x) = (1 + |x|^2)^{-\sigma/2}$ for some $\sigma \in [0, \infty)$. Then, for $m \in \mathbb{Z}$ and $p \in [1, \infty]$, we have*

$$C^{-1} \|f\|_{B^m_{p,\infty}(\nu)} \leq \|f\|_{W^{m,p}(\nu)} \leq C \|f\|_{B^m_{p,1}(\nu)}$$

where C is a (positive) constant depending on σ, m and p, $\|f\|_{W^{m,p}(\nu)}$ is the weighted Sobolev norm defined by

$$\|f\|_{W^{m,p}(\nu)} := \sum_{k=0}^{m} \|\nabla^k f\|_{L^p(\nu)}, \quad m \in \mathbb{N} \cup \{0\}$$

$$\|f\|_{W^{-m,p}(\nu)} := \sup_{g \in \mathcal{S}(\mathbb{R}^3); g \neq 0} \frac{|{}_{\mathcal{S}'}\langle f, g \rangle_{\mathcal{S}}|}{\|g\|_{W^{m,p^*}(\nu)}}, \quad m \in \mathbb{N},$$

and p^* is the Hölder conjugate of p.

PROOF. It is sufficient to prove the result for the case that $m = 0$, because of the effect of derivatives (see [**MW17c**, Proposition 3]) for $m \in \mathbb{N}$ and the duality of Besov spaces (see Proposition 2.4(ii)). The triangle inequality implies

$$\|f\|_{L^p(\nu)} \leq \sum_{j=-1}^{\infty} \|\Delta_j f\|_{L^p(\nu)} = \|f\|_{B^0_{p,1}(\nu)}.$$

Hence, we have the second inequality of the assertion. Let $\varphi_j := \varphi(2^{-j}\cdot)$ for $j \in \mathbb{N} \cup \{0\}$, $\varphi_{-1} := \chi$ and $\varphi_{-2} = 0$. The fact that

$$\nu(x) \leq 2^{\sigma/2}(1 + |x - y|^2)^{\sigma/2} \nu(y) \quad x, y \in \mathbb{R}^3 \tag{2.1}$$

implies

$$\begin{aligned}
&\|f\|_{B^0_{p,\infty}(\nu)} \\
&= \sup_{j\in\mathbb{N}\cup\{-1,0\}} \|\Delta_j f\|_{L^p(\nu)} \\
&= \sup_{j\in\mathbb{N}\cup\{-1,0\}} \left(\int_{\mathbb{R}^3}\left(\int_{\mathbb{R}^3}\left(\mathcal{F}^{-1}\varphi_j\right)(x-y)f(y)dy\right)^p \nu(x)dx\right)^{1/p} \\
&\le 2^{\sigma/(2p)} \sup_{j\in\mathbb{N}\cup\{-1,0\}} \left(\int_{\mathbb{R}^3}\left(\int_{\mathbb{R}^3}\left(1+|x-y|^2\right)^{\sigma/2p}\left(\mathcal{F}^{-1}\varphi_j\right)\right.\right. \\
&\qquad\qquad \left.\left.(x-y)f(y)\nu(y)^{1/p}dy\right)^p dx\right)^{1/p}.
\end{aligned}$$

Hence, by Young's inequality we have

$$\|f\|_{B^0_{p,\infty}(\nu)} \le 2^{\sigma/(2p)} \sup_{j\in\mathbb{N}\cup\{-1,0\}} \left(\int_{\mathbb{R}^3}(1+|x|^2)^{\sigma/2p}|(\mathcal{F}^{-1}\varphi_j)(x)|dx\right)\|f\|_{L^p(\nu)}.$$

In view of this inequality, it is sufficient to show that

$$\sup_{j\in\mathbb{N}\cup\{-1,0\}} \int_{\mathbb{R}^3}(1+|x|^2)^{\sigma/2p}|(\mathcal{F}^{-1}\varphi_j)(x)|dx < \infty.$$

However, this claim is a special case ($\alpha=\beta$ and $s=1$) of (2.4) in the proof of Proposition 2.4 below. So, we omit it. □

Let us now make, for simplicity, the convention that all constants C appearing in estimates (sometimes with indices indicating their dependence on parameters) are positive.

COROLLARY 2.3. Assume that $\nu(x)=(1+|x|^2)^{-\sigma/2}$ for some $\sigma\in[0,\infty)$, and define the Sobolev space $W^{s,p}(\nu)$ for $s\in\mathbb{R}\setminus\mathbb{Z}$ by the interpolation spaces of $W^{m,p}(\nu)$ with $m\in\mathbb{Z}$. Then, for $s\in\mathbb{R}$ and $p\in[1,\infty]$, we have

$$C^{-1}\|f\|_{B^s_{p,\infty}(\nu)} \le \|f\|_{W^{s,p}(\nu)} \le C\|f\|_{B^s_{p,1}(\nu)}$$

where C is a constant depending on σ, s and p.

PROOF. Since $\{B^s_{p,r}; s\in\mathbb{R}\}$ is the series of interpolation spaces, the assertion follows from the interpolation inequalities and Proposition 2.2. □

PROPOSITION 2.4.

(i) For $s_1,s_2\in\mathbb{R}$, $\theta\in(0,1)$ and $p,r\in[1,\infty]$

$$\|f\|_{B^{\theta s_1+(1-\theta)s_2}_{p,r}(\nu)} \le \|f\|^{\theta}_{B^{s_1}_{p,r}(\nu)}\|f\|^{1-\theta}_{B^{s_2}_{p,r}(\nu)}, \quad f\in B^{s_1}_{p,r}(\nu)\cap B^{s_2}_{p,r}(\nu).$$

(ii) For $s\in\mathbb{R}$ and $p_1,p_2,r_1,r_2\in[1,\infty]$ such that $1=1/p_1+1/p_2$ and $1=1/r_1+1/r_2$, there exists a constant C depending on s, p_1, p_2, r_1 and r_2 satisfying

$$\left|\int_{\mathbb{R}^3} fg\nu dx\right| \le C\|f\|_{B^{-s}_{p_1,r_2}(\nu)}\|g\|_{B^s_{p_2,r_2}(\nu)}, \quad f\in B^{-s}_{p_1,r_2}(\nu),\ g\in B^s_{p_2,r_2}(\nu).$$

(iii) Assume that $\nu(x)=(1+|x|^2)^{-\sigma/2}$ for some $\sigma\in[0,\infty)$. For $\alpha\in\mathbb{R}$, $\beta\in[0,\infty)$, and $p,r\in[1,\infty]$, there exists a constant C depending on α, β, p and r such that

$$\|e^{t\triangle}\Delta_j f\|_{B^{\alpha}_{p,r}(\nu)} \le C(2^{2j}t)^{-\beta}\|\Delta_j f\|_{B^{\alpha}_{p,r}(\nu)}$$

for $j \in \mathbb{N} \cup \{0\}$, $f \in B^{\alpha}_{p,r}(\nu)$, where $\{e^{t\triangle}; t \geq 0\}$ is the heat semigroup generated by $\triangle$ on $\mathbb{R}^3$. In particular,

$$\|e^{t\triangle} f\|_{B^{\alpha}_{p,r}(\nu)} \leq C(1 + t^{-\beta})\|f\|_{B^{\alpha-2\beta}_{p,r}(\nu)}, \quad f \in B^{\alpha-2\beta}_{p,r}(\nu).$$

(iv) Assume that $\nu(x) = (1 + |x|^2)^{-\sigma/2}$ for some $\sigma \in [0, \infty)$. For $\alpha, \beta \in \mathbb{R}$, $p, q \in [1, \infty)$ and $r \in [1, \infty]$ such that

$$\alpha < \beta, \quad p > q, \quad \beta - \alpha = 3\left(\frac{1}{q} - \frac{1}{p}\right),$$

there exists a constant C depending on σ, α, β, p and q such that

$$\|f\|_{B^{\alpha}_{p,r}(\nu)} \leq C\|f\|_{B^{\beta}_{q,r}(\nu^{q/p})}, \quad f \in B^{\beta}_{p,r}(\nu^{q/p}).$$

Proof. The proofs for (i), (ii) and (iii) are similar to the case without weights. See [**BCD11**, Theorem 2.80 and Proposition 2.76] for (i) and (ii).

For the first assertion of (iii) it is sufficient to prove the case where $r = 1$. The proof of (iii) is similar to [**MW17c**, Lemma 3 and Proposition 5]. First we show that there exists $\varepsilon \in (0, 1)$ such that

$$\|e^{t\triangle}\Delta_j f\|_{L^p(\nu)} \leq Ce^{-\varepsilon 2^{2j}t}\|\Delta_j f\|_{L^p(\nu)} \tag{2.2}$$

for $j \in \mathbb{N} \cup \{0\}$. Set φ_j for $j \in \mathbb{N} \cup \{-1, 0\}$ as in the proof of Proposition 2.2. The fact that $\Delta_i\Delta_j = 0$ for $|i - j| \geq 2$ and (2.1) imply

$$\begin{aligned}
&\|e^{t\triangle}\Delta_j f\|_{L^p(\nu)} \\
&\leq C \sum_{i=j-1}^{j+1} \|\Delta_i\Delta_j e^{t\triangle} f\|_{L^p(\nu)} \\
&\leq C \sum_{i=j-1}^{j+1} \left(\int_{\mathbb{R}^3} \left|\int_{\mathbb{R}^3} (\mathcal{F}^{-1}\varphi_i)(x-y)e^{-|x-y|^2/(4t)}(\Delta_j f)(y)dy\right|^p \nu(x)dx\right)^{1/p} \\
&\leq C2^{\sigma/2p} \sum_{i=j-1}^{j+1} \left(\int_{\mathbb{R}^3} \left|\int_{\mathbb{R}^3} (1 + |x-y|^2)^{\sigma/2p}\left[\mathcal{F}^{-1}\left(e^{-2t|\cdot|^2}\varphi_i\right)\right](x-y)\right.\right. \\
&\qquad\qquad \left.\left. \times (\Delta_j f)(y)\nu(y)^{1/p}dy\right|^p dx\right)^{1/p}.
\end{aligned}$$

Hence, by Young's inequality we have

$$\begin{aligned}
&\|e^{t\triangle}\Delta_j f\|_{L^p(\nu)} \\
&\leq C\|\Delta_j f\|_{L^p(\nu)} \sum_{i=j-1}^{j+1} \int_{\mathbb{R}^3} \left|(1 + |y|^2)^{\sigma/2p}\left[\mathcal{F}^{-1}\left(e^{-2t|\cdot|^2}\varphi_i\right)\right](y)\right| dy.
\end{aligned}$$

In view of this inequality, it is sufficient to show

$$\max_{i=j,j\pm1} \int_{\mathbb{R}^3} \left|(1 + |y|^2)^{\sigma/2p}\left[\mathcal{F}^{-1}\left(e^{-2t|\cdot|^2}\varphi_i\right)\right](y)\right| dy \leq Ce^{-\varepsilon 2^{2j}t}.$$

for some $\varepsilon \in (0, 1)$. The proof of this estimate is similar to that of [**BCD11**, Lemma 2.4]. So, we omit it.

Next we prove

$$\|\Delta_i f\|_{L^p(\nu)} \leq C\|f\|_{B^{\alpha-2\beta}_{p,r}(\nu)}, \quad f \in B^{\alpha-2\beta}_{p,r}(\nu),\ i \in \{-1, 0\}. \tag{2.3}$$

For $m \in \mathbb{N}$,

$$\begin{aligned}
&\|\Delta_i f\|_{L^p(\nu)}^p \\
&= \int_{\mathbb{R}^3} \left| \int_{\mathbb{R}^3} \left[(1-\triangle)^m \mathcal{F}^{-1}\varphi_i\right](x-y) \cdot \left[(1-\triangle)^{-m} f\right](y)dy \right|^p \nu(x)dx \\
&\leq 2^{\sigma/2} \int_{\mathbb{R}^3} \left| \int_{\mathbb{R}^3} (1+|x-y|^2)^{\sigma/2p} \left[\mathcal{F}^{-1}\left((1+|\cdot|^2)^m \varphi_i\right)\right](x-y) \right. \\
&\qquad\qquad \left. \times \left[(1-\triangle)^{-m} f\right](y)\nu(y)^{1/p} dy \right|^p dx.
\end{aligned}$$

Hence, by Young's inequality we have

$$\begin{aligned}
&\|\Delta_i f\|_{L^p(\nu)}^p \\
&\leq C \left\|(1-\triangle)^{-m} f\right\|_{L^p(\nu)}^p \left(\int_{\mathbb{R}^3} (1+|y|^2)^{\sigma/2p} \left| \left[\mathcal{F}^{-1}\left((1+|\cdot|^2)^m \varphi_i\right)\right](y) \right| dy \right)^p.
\end{aligned}$$

Since $i = -1, 0$ and $(1+|\cdot|^2)^m \varphi_i \in \mathcal{S}(\mathbb{R}^3)$, we obtain

$$\|\Delta_i f\|_{L^p(\nu)} \leq C \left\|(1-\triangle)^{-m} f\right\|_{L^p(\nu)}.$$

In view of Proposition 2.2, by taking $2m > |\alpha - 2\beta|$, we have (2.3).

Now we prove (iii) by applying (2.2). From (2.2) and the fact that $\Delta_i \Delta_j = 0$ for $|i-j| \geq 2$, we have that for $j \in \mathbb{N} \cup \{0\}$ and $i \in \mathbb{N} \cup \{-1, 0\}$

$$\begin{aligned}
&2^{\alpha i} \|e^{t\triangle} \Delta_i \Delta_j f\|_{L^p(\nu)} \\
&\leq C 2^{\alpha i} e^{-\varepsilon 2^{2j} t} \|\Delta_i \Delta_j f\|_{L^p(\nu)} \\
&\leq C 2^{\alpha i} (2^{2j} t)^{\beta} e^{-\varepsilon 2^{2j} t} (2^{2j} t)^{-\beta} \|\Delta_i \Delta_j f\|_{L^p(\nu)} \\
&\leq C 2^{\alpha i} (2^{2j} t)^{-\beta} \|\Delta_i \Delta_j f\|_{L^p(\nu)}.
\end{aligned}$$

The first assertion follows from this inequality and (2.3). Since (2.2) also implies that for $j \in \mathbb{N} \cup \{0\}$

$$\begin{aligned}
2^{\alpha j} \|e^{t\triangle} \Delta_j f\|_{L^p(\nu)} &\leq C t^{-\beta} 2^{(\alpha - 2\beta) j} (2^{2j} t)^{\beta} e^{-\varepsilon 2^{2j} t} \|\Delta_j f\|_{L^p(\nu)} \\
&\leq C t^{-\beta} 2^{(\alpha - 2\beta) j} \|\Delta_j f\|_{L^p(\nu)},
\end{aligned}$$

by (2.3) we get the second assertion.

Next we prove (iv). We remark that the proof is similar to that of [**MW17c**, Proposition 3.7]. The fact that $\Delta_i \Delta_j = 0$ for $|i-j| \geq 2$, together with (2.1) implies

$$\begin{aligned}
&\|\Delta_j f\|_{L^p(\nu)} \\
&= \left(\int_{\mathbb{R}^3} \left| \int_{\mathbb{R}^3} (\mathcal{F}^{-1}\varphi_j)(x-y) [(\Delta_{j-1} + \Delta_j + \Delta_{j+1}) f](y) dy \right|^p \nu(x) dx \right)^{1/p} \\
&\leq 2^{\sigma/2p} \left(\int_{\mathbb{R}^3} \left| \int_{\mathbb{R}^3} (1+|x-y|^2)^{\sigma/2p} (\mathcal{F}^{-1}\varphi_j)(x-y) \right. \right. \\
&\qquad\qquad \left. \left. \times [(\Delta_{j-1} + \Delta_j + \Delta_{j+1}) f](y) \nu(y)^{1/p} dy \right|^p dx \right)^{1/p}.
\end{aligned}$$

Hence, by letting $s \in [1, \infty]$ such that $1/s + 1/q = 1/p + 1$, from the Young's inequality we have

$$\|\Delta_j f\|_{L^p(\nu)} \leq 2^{\sigma/2p} \left(\int_{\mathbb{R}^3} (1 + |x|^2)^{\sigma s/2p} |(\mathcal{F}^{-1}\varphi_j)(x)|^s dx \right)^{1/s} \times \left\| \nu^{1/p} (\Delta_{j-1} + \Delta_j + \Delta_{j+1}) f \right\|_{L^q(1)}.$$

Thus, we have

$$\|f\|_{B^\alpha_{p,r}(\nu)} \leq C \left(\sup_{j \in \mathbb{N} \cup \{-1,0\}} 2^{js(\alpha-\beta)} \int_{\mathbb{R}^3} (1 + |x|^2)^{\sigma s/2p} |(\mathcal{F}^{-1}\varphi_j)(x)|^s dx \right)^{1/s} \|f\|_{B^\beta_{q,r}(\nu^{q/p})}.$$

In view of this inequality, it is sufficient to show

$$\sup_{j \in \mathbb{N} \cup \{-1,0\}} 2^{js(\alpha-\beta)} \int_{\mathbb{R}^3} (1 + |x|^2)^{\sigma s/2p} |(\mathcal{F}^{-1}\varphi_j)(x)|^s dx < \infty. \tag{2.4}$$

Since

$$(\mathcal{F}^{-1}\varphi_j)(x) = 2^{3j} (\mathcal{F}^{-1}\varphi)(2^j x), \quad x \in \mathbb{R}^3 \tag{2.5}$$

and $\mathcal{F}^{-1}\varphi \in \mathcal{S}(\mathbb{R}^3)$, we have

$$\begin{aligned} &\sup_{j \in \mathbb{N} \cup \{0\}} 2^{js(\alpha-\beta)} \int_{\mathbb{R}^3} (1 + |x|^2)^{\sigma s/2p} |(\mathcal{F}^{-1}\varphi_j)(x)|^s dx \\ &= \sup_{j \in \mathbb{N} \cup \{0\}} 2^{js(\alpha-\beta)} 2^{3j(s-1)} \int_{\mathbb{R}^3} (1 + 2^{-2j}|x|^2)^{\sigma s/2p} |(\mathcal{F}^{-1}\varphi)(x)| dx. \end{aligned}$$

The assumption in (iv) implies that $s(\alpha - \beta) + 3(s-1) = 0$ and the finiteness for the case that $j = -1$ is trivial. Therefore, we obtain (2.4). □

Next we prepare the notations and estimates of paraproducts by following [**BCD11**, Chapter 2]. Let

$$S_j f := \sum_{k=-1}^{j-1} \Delta_k f, \quad j \in \mathbb{N} \cup \{0\}.$$

For simplicity of notation, let $\Delta_{-2} f := 0$ and $S_{-1} f := 0$. We define

$$\begin{aligned} f \varolessthan g &:= \sum_{j=0}^{\infty} (S_j f) \Delta_{j+1} g \\ f \circleddash g &:= \sum_{j=-1}^{\infty} \Delta_j f \, (\Delta_{j-1} g + \Delta_j g + \Delta_{j+1} g) \\ f \varogreaterthan g &:= g \varolessthan f \end{aligned}$$

By the definitions of $\{\Delta_j\}$, $\{S_j\}$, $\varolessthan$, $\circleddash$, and $\varogreaterthan$, we have a decomposition

$$fg = f \varolessthan g + f \circleddash g + f \varogreaterthan g,$$

which is called the Bony decomposition. Let $f \mathbin{\underline{\varolessthan}} g := f \varolessthan g + f \circleddash g$ and $f \mathbin{\underline{\varogreaterthan}} g := f \varogreaterthan g + f \circleddash g$.

PROPOSITION 2.5.

(i) For $s \in \mathbb{R}$ and $p, p_1, p_2, r \in [1,\infty]$ such that $1/p = 1/p_1 + 1/p_2$,

$$\|f \circledless g\|_{B^s_{p,r}(\nu)} \le C\|f\|_{L^{p_1}(\nu)}\|g\|_{B^s_{p_2,r}(\nu)}, \qquad f \in L^{p_1}(\nu),\ g \in B^s_{p_2,r}(\nu),$$

where C is a constant depending on s, p_1, p_2 and r.

(ii) For $s \in \mathbb{R}$, $t \in (-\infty, 0)$, and $p, p_1, p_2, r_1, r_2 \in [1,\infty]$ such that

$$\frac{1}{p} = \frac{1}{p_1} + \frac{1}{p_2} \quad \text{and} \quad \frac{1}{r} = \min\left\{1, \frac{1}{r_1} + \frac{1}{r_2}\right\},$$

$$\|f \circledless g\|_{B^{s+t}_{p,r}(\nu)} \le C\|f\|_{B^t_{p_1,r_1}(\nu)}\|g\|_{B^s_{p_2,r_2}(\nu)}, \qquad f \in B^t_{p_1,r_1}(\nu),\ g \in B^s_{p_2,r_2}(\nu),$$

where C is a constant depending on s, p_1, p_2, r_1 and r_2.

(iii) For $s_1, s_2 \in \mathbb{R}$ such that $s_1 + s_2 > 0$, and $p, p_1, p_2, r, r_1, r_2 \in [1,\infty]$ such that

$$\frac{1}{p} = \frac{1}{p_1} + \frac{1}{p_2} \quad \text{and} \quad \frac{1}{r} = \frac{1}{r_1} + \frac{1}{r_2},$$

it holds that

$$\|f \circleddash g\|_{B^{s_1+s_2}_{p,r}(\nu)} \le C\|f\|_{B^{s_1}_{p_1,r_1}(\nu)}\|g\|_{B^{s_2}_{p_2,r_2}(\nu)}, \qquad f \in B^{s_1}_{p_1,r_1}(\nu),\ g \in B^{s_2}_{p_2,r_2}(\nu),$$

where C is a constant depending on s_1, s_2, p_1, p_2, r_1 and r_2.

(iv) For $s \in (-\infty, 0)$ and $t \in (0,\infty)$ such that $s+t > 0$, and $p, p_1, p_2, r, r_1, r_2 \in [1,\infty]$ such that

$$\frac{1}{p} = \frac{1}{p_1} + \frac{1}{p_2} \quad \text{and} \quad \frac{1}{r} = \frac{1}{r_1} + \frac{1}{r_2},$$

it holds that

$$\|fg\|_{B^s_{p,r}(\nu)} \le C\|f\|_{B^s_{p_1,r_1}(\nu)}\|g\|_{B^t_{p_2,r_2}(\nu)}, \qquad f \in B^s_{p_1,r_1}(\nu),\ g \in B^t_{p_2,r_2}(\nu),$$
$$\|fg\|_{B^t_{p,r}(\nu)} \le C\|f\|_{B^t_{p_1,r_1}(\nu)}\|g\|_{B^t_{p_2,r_2}(\nu)}, \qquad f \in B^t_{p_1,r_1}(\nu),\ g \in B^t_{p_2,r_2}(\nu),$$

where C is a constant depending on s, t, p_1, p_2, r_1 and r_2.

(v) For $s \in (0,\infty)$, $\varepsilon \in (0,1)$ and $p, p_1, p_2, r \in [1,\infty]$ such that $1/p = 1/p_1 + 1/p_2$,

$$\|f^2\|_{B^s_{p,r}(\nu)} \le C\|f\|_{L^{p_1}(\nu)}\|f\|_{B^{s+\varepsilon}_{p_2,r}(\nu)}, \qquad f \in L^{p_1}(\nu) \cap B^{s+\varepsilon}_{p_2,r}(\nu)$$

where C is a constant depending on s, ε, p_1, p_2 and r.

(vi) For $s \in (0,\infty)$, $\varepsilon \in (0,1)$ and $p, r \in [1,\infty]$,

$$\|f^3\|_{B^s_{p,r}(\nu)} \le C\|f\|^2_{L^{4p}(\nu)}\|f\|_{B^{s+\varepsilon}_{2p,r}(\nu)}, \qquad f \in L^{4p}(\nu) \cap B^{s+\varepsilon}_{2p,r}(\nu)$$

where C is a positive constant depending on s, ε, p and r.

PROOF. For (i) and (ii), see [**MW17b**, Proposition A.7]. The proofs of (iii) are similar to the case of the Besov spaces without weights (see [**BCD11**, Theorem 2.85]). From (i), (ii) and (iii), (iv) follows . The proofs of (v) and (vi) are also similar to those of [**AK20**, Proposition 2.1 (v) and (vi)], respectively. □

PROPOSITION 2.6. Assume that $\nu(x) = (1 + |x|^2)^{-\sigma/2}$ for some $\sigma \in (3,\infty)$. Then, for $\alpha, \beta \in \mathbb{R}$ such that $\alpha < \beta$, and $p, q, r \in [1,\infty]$ such that $p < q$, the embedding $B^\beta_{q,r}(\nu) \hookrightarrow B^\alpha_{p,r}(\nu)$ is compact.

PROOF. For a bounded sequence $\{f_n\} \subset B^{\beta}_{q,r}(\nu)$, by considering

$$\{(1-\triangle)^{-(\alpha+\beta)/4} f_n\}$$

instead of $\{f_n\}$, we are able to assume $\alpha < 0 < \beta$. Hence, in view of the fact that $B^s_{p,\infty}(\nu) \subset B^{s'}_{p,1}(\nu)$ for $p \in [1,\infty]$ and $s, s' \in \mathbb{R}$ such that $s' < s$, and Proposition 2.2, it is sufficient to show the compactness of the embedding $L^q(\nu) \hookrightarrow B^{-\alpha}_{p,1}(\nu)$ for $\alpha \in (0,1)$ and $p, q \in [1,\infty]$ such that $p < q$.

Let $\varphi_j := \varphi(2^{-j}\cdot)$ for $j \in \mathbb{N} \cup \{0\}$ and $\varphi_{-1} := \chi$, where φ and χ are functions prepared for the definition for Besov spaces (see the beginning of Section 2.1). Let $\{f_n\}$ be a bounded sequence in $L^p(\nu)$. Since φ_i has a compact support for each i, then for each $i, j \in \mathbb{N} \cup \{-1, 0\}$ and $m \in \mathbb{N}$ we have, for some constant $C_{m,i}$

$$\begin{aligned}
\|\varphi_i \Delta_j f_n\|_{W^{m,p}} &= \sum_{k \in (\mathbb{N}\cup\{0\})^3; |k| \le m} \left\|\partial^k (\varphi_i \Delta_j f_n)\right\|_{L^p} \\
&= \sum_{k,l \in (\mathbb{N}\cup\{0\})^3; |k|+|l| \le m} \left\|(\partial^k \varphi_i)(\partial^l \Delta_j f_n)\right\|_{L^p} \\
&\le C_{m,i} \sum_{l \in (\mathbb{N}\cup\{0\})^3; |l| \le m} \left\|(\partial^l \mathcal{F}^{-1}\varphi_j) * f_n\right\|_{L^p(\nu)}.
\end{aligned}$$

This and Hölder's inequality imply

$$\|\varphi_i \Delta_j f_n\|_{W^{m,p}} \le C_{m,i,j} \|f_n\|_{L^p(\nu)}$$

for some constant $C_{m,i,j}$. Hence, for each $i, j \in \mathbb{N}\cup\{-1,0\}$ and $m \in \mathbb{N}$, $\{\varphi_i \Delta_j f_n; n \in \mathbb{N}\}$ are uniformly bounded in $W^{m,p}$. In view of the compactness of the support of φ_i, Sobolev's inequality and the Ascoli-Arzelà theorem yield that for each $i, j \in \mathbb{N}\cup\{-1,0\}$, there exists a subsequence $\{f_{n(k)}\}$ of $\{f_n\}$ such that $\{\varphi_i \Delta_j f_{n(k)}\}$ converges in $C_b(\mathbb{R}^3)$ with the topology of uniform convergence. By the diagonal method, we have a subsequence $\{f_{n(k)}\}$ such that $\{\varphi_i \Delta_j f_{n(k)}\}$ converges in $C_b(\mathbb{R}^3)$ for all $i, j \in \mathbb{N}\cup\{-1,0\}$. Let us take $a \in [1,\infty)$ satisfying $(1/a) + (1/q) = 1/p$. For $k, l \in \mathbb{N}$ and $N \in \mathbb{N}$ we have

$$\begin{aligned}
&\|f_{n(k)} - f_{n(l)}\|_{B^{-\alpha}_{p,1}(\nu)} \\
&\le \sum_{j=-1}^{N} 2^{-j\alpha} \|\Delta_j (f_{n(k)} - f_{n(l)})\|_{L^p(\nu)} + \sum_{j=N+1}^{\infty} 2^{-j\alpha} \left\|\Delta_j (f_{n(k)} - f_{n(l)})\right\|_{L^p(\nu)} \\
&\le \sum_{i,j=-1}^{N} 2^{-j\alpha} \|\varphi_i \Delta_j (f_{n(k)} - f_{n(l)})\|_{L^p(\nu)} \\
&\quad + \sum_{j=-1}^{N} 2^{-j\alpha} \left\| \left(1 - \sum_{i=-1}^{N} \varphi_i\right) \Delta_j (f_{n(k)} - f_{n(l)}) \right\|_{L^p(\nu)} \\
&\quad + 2^{-\alpha(N+1)/2} \sum_{j=N+1}^{\infty} 2^{-j\alpha/2} \|\Delta_j (f_{n(k)} - f_{n(l)})\|_{L^p(\nu)} \\
&\le \sum_{i,j=-1}^{N} 2^{-j\alpha} \|\varphi_i \Delta_j (f_{n(k)} - f_{n(l)})\|_{L^p(\nu)}
\end{aligned}$$

$$
\begin{aligned}
&+ \left\| 1 - \sum_{i=-1}^{N} \varphi_i \right\|_{L^a(\nu)} \sum_{j=-1}^{N} 2^{-j\alpha} \left\| \Delta_j (f_{n(k)} - f_{n(l)}) \right\|_{L^q(\nu)} \\
&+ 2^{-\alpha(N+1)/2} \| f_{n(k)} - f_{n(l)} \|_{B^{-\alpha}_{p,1}(\nu)} \\
\leq & \sum_{i,j=-1}^{N} 2^{-j\alpha} \| \varphi_i \Delta_j (f_{n(k)} - f_{n(l)}) \|_{L^p(\nu)} \\
&+ \left(2^{-\alpha(N+1)/2} + \left\| \mathbb{I}_{[3/4,\infty)} \left(2^{-N} | \cdot | \right) \right\|_{L^a(\nu)} \right) \| f_{n(k)} - f_{n(l)} \|_{B^{-\alpha}_{q,1}(\nu)}.
\end{aligned}
$$

Since

$$
\| f_{n(k)} - f_{n(l)} \|_{B^{-\alpha}_{q,1}(\nu)} \leq \| f_{n(k)} - f_{n(l)} \|_{L^q(\nu)} \leq 2 \sup_{n \in \mathbb{N}} \| f_n \|_{L^q(\nu)} < \infty,
$$

for any $\varepsilon \in (0,1)$ there exists a sufficiently large N such that

$$
\| f_{n(k)} - f_{n(l)} \|_{B^{-\alpha}_{p,1}(\nu)} \leq \sum_{i,j=-1}^{N} 2^{-j\alpha} \| \varphi_i \Delta_j f_{n(k)} - \varphi_i \Delta_j f_{n(l)} \|_{L^p(\nu)} + \varepsilon.
$$

This estimate and the fact that $\{\varphi_i \Delta_j f_{n(k)}\}$ converges in $C_b(\mathbb{R}^3)$ for all $i, j \in \mathbb{N} \cup \{-1, 0\}$ yield the conclusion. □

LEMMA 2.7. *Assume that the total mass of the measure $\nu(x)dx$ is finite. Then:*

(i) *For $\theta \in (0, 9/16)$ there exists a (positive) constant C depending on θ such that for $\delta \in (0,1]$ and $f \in L^4(\nu) \cap B_2^{15/16}(\nu)$*

$$
\| f \|_{B_2^\theta(\nu)} \leq \delta \left(\| f \|^4_{L^4(\nu)} + \| f \|^2_{B_2^{15/16}(\nu)} \right)^{7/8} + C \delta^{-16/19}.
$$

(ii) *For $\theta \in (0, 9/16)$ there exists a (positive) constant C depending on θ such that for $\delta \in (0,1]$ and $f \in L^4(\nu) \cap B_2^{15/16}(\nu)$*

$$
\| f^2 \|_{B_{4/3}^\theta(\nu)} \leq \delta \left(\| f \|^4_{L^4(\nu)} + \| f \|^2_{B_2^{15/16}(\nu)} \right)^{7/8} + C \delta^{-26/9}.
$$

(iii) *For $\theta \in (0, 9/16)$ there exists a (positive) constant C depending on θ such that for $\delta \in (0,1]$ and $f \in L^4(\nu) \cap B_2^{15/16}(\nu)$*

$$
\| f^3 \|_{B_1^\theta(\nu)} \leq \delta \left(\| f \|^4_{L^4(\nu)} + \| f \|^2_{B_2^{15/16}(\nu)} \right) + C \delta^{-10}.
$$

PROOF. The proofs are similar to those in [**AK20**, Lemma 2.2]. So, we omit them. □

2.2. Approximation operators and commutator estimates

We define ψ, ρ, ψ_N, ρ_M, P_N, $P_{M,N}$ and $P^*_{M,N}$ as in Section 1.2. As remarked in Section 1.2, $P^*_{M,N}$ is the dual operator of $P_{M,N}$ with respect to $L^2(dx)$. In the rest of the paper, we fix ψ and ρ and do not mention explicitly the dependence on them, even if the constants that will appear in the estimates below are depending on them.

For simplicity of notation, we denote $B^s_{p,\infty}(\nu)$ by $B^s_p(\nu)$ for $s \in \mathbb{R}$ and $p \in [1, \infty]$. Moreover, when $\nu = 1$, we denote $L^p(\nu)$, $B^s_{p,r}(\nu)$ and $B^s_p(\nu)$ simply by L^p, $B^s_{p,r}$ and B^s_p, respectively.

Proposition 2.8 (cf. [**Ste70**, Theorem 3 of Chapter IV]). For $p \in (1,\infty)$, there exists a constant C independent of N such that

$$\|P_N f\|_{L^p} \le C\|f\|_{L^p}, \quad f \in L^p.$$

Proof. This proposition immediately follows from [**Ste70**, Theorem 3 of Chapter IV] and the comment just after the theorem. □

Lemma 2.9. *Let $\alpha \in \mathbb{R}$, $\beta \in (0,\infty)$ and $p \in [1,\infty]$. Assume that $\alpha \neq \beta$. Then, there exists an absolute constant C such that*

$$\|P_N f\|_{B_p^\alpha(\nu)} \le C2^{\beta N}\|P_N f\|_{B_p^{\alpha-\beta}(\nu)}$$

for $f \in \mathcal{S}(\mathbb{R}^3)$ and $N \in \mathbb{N}$. In particular,

$$\|P_N f\|_{B_p^\alpha} \le C2^{\beta N}\|P_N f\|_{B_p^{\alpha-\beta}},$$
$$\left\|P_{M,N}^* f\right\|_{B_p^\alpha(\nu)} \le C2^{\beta N}\|P_{M,N}^* f\|_{B_p^{\alpha-\beta}(\nu)}$$

for $f \in \mathcal{S}(\mathbb{R}^3)$ and $M, N \in \mathbb{N}$.

Proof. In view of the supports of φ and ψ_N,

$$\Delta_j P_N = 0, \quad \text{if } 2^N \le \frac{\sqrt{3}}{8}2^j. \tag{2.6}$$

Hence, we have

$$\|P_N f\|_{B_p^\alpha(\nu)} \le C2^{\beta N} \sup_{j\in\mathbb{N}\cup\{-1,0\};2^j\le 2^{N+3}/\sqrt{3}} 2^{j(\alpha-\beta)}\|\Delta_j P_N f\|_{L^p(\nu)},$$

from which the assertion follows. □

Lemma 2.10. *Let $\alpha \in \mathbb{R}$, $\beta \in [0,\infty)$ and $p \in (1,\infty)$. Then, there exists a constant C depending on β and p such that*

$$\|P_N f - f\|_{B_p^\alpha} \le C2^{-\beta N}\|f\|_{B_p^{\alpha+\beta}}$$

for $f \in \mathcal{S}(\mathbb{R}^3)$ and $N \in \mathbb{N}$. In particular,

$$\|P_N f\|_{B_p^\alpha} \le C\|f\|_{B_p^\alpha}$$

for $f \in \mathcal{S}(\mathbb{R}^3)$ and $N \in \mathbb{N}$.

Proof. In view of the supports of φ and ψ_N, we have

$$\Delta_j(I - P_N) = 0, \quad \text{if } 2^N \ge \frac{3\sqrt{3}}{4}2^j.$$

Hence, the commutativity of Δ_j and $I - P_N$ implies

$$\begin{aligned}
&\|(I-P_N)f\|_{B_p^\alpha} \\
&= \sup_{j\in\mathbb{N}\cup\{-1,0\};2^j\ge 4N/(3\sqrt{3})} 2^{j\alpha}\|(I-P_N)\Delta_j f\|_{L^p} \\
&\le \left(\frac{3\sqrt{3}}{4}\right)^\beta 2^{-\beta N} \sup_{j\in\mathbb{N}\cup\{-1,0\};2^j\ge 2^{N+2}/(3\sqrt{3})} 2^{j(\alpha+\beta)}\|(I-P_N)\Delta_j f\|_{L^p}.
\end{aligned}$$

In view of this inequality and Proposition 2.8, there exists a constant C_p depending on p, such that

$$\|P_N f - f\|_{B_p^\alpha} \leq C_p \left(\frac{3\sqrt{3}}{4}\right)^\beta 2^{-\beta N} \|f\|_{B_p^{\alpha+\beta}}.$$

This yields the conclusion. ☐

LEMMA 2.11. *Let $\alpha \in \mathbb{R}$, $\beta \in (0,\infty)$, $p \in (1,\infty)$, and assume that $\nu(x) = (1+|x|)^{-\sigma}$ $(x \in \mathbb{R}^3)$ for some $\sigma \in [0,\infty)$. Then, for $M \in \mathbb{N}$ there exists a constant C depending on σ, α, β, p, ψ and ρ such that*

$$\left\|P_{M,N}^* f - \rho_M f\right\|_{B_p^\alpha(\nu)} \leq CM^\sigma 2^{-\beta N} \|f\|_{B_p^{\alpha+\beta}(\nu)}$$

for $f \in \mathcal{S}(\mathbb{R}^3)$ and $N \in \mathbb{N}$. In particular,

$$\left\|P_{M,N}^* f\right\|_{B_p^\alpha(\nu)} \leq C\left(1 + M^\sigma 2^{-\beta N}\right) \|f\|_{B_p^{\alpha+\beta}(\nu)}$$

for $f \in \mathcal{S}(\mathbb{R}^3)$ and $N \in \mathbb{N}$.

PROOF. By Lemma 2.10 and the fact that $\nu \leq 1$,

$$\|P_N(\rho_M f) - \rho_M f\|_{B_p^\alpha(\nu)} \leq C_p 2^{-\beta N} \|\rho_M f\|_{B_p^{\alpha+\beta}}.$$

Hence, it is sufficient to prove

$$\|\rho_M f\|_{B_p^{\alpha+\beta}} \leq CM^\sigma \|f\|_{B_p^{\alpha+\beta}(\nu)}. \tag{2.7}$$

Let q be the Hölder conjugate of p. [**BCD11**, Proposition 2.76] and Proposition 2.4 imply

$$\begin{aligned}
\|\rho_M f\|_{B_p^{\alpha+\beta}} &\leq C \sup_{h \in \mathcal{S}(\mathbb{R}^3);\ \|h\|_{B_{q,1}^{-(\alpha+\beta)}} \leq 1} \langle h, \rho_M f\rangle \\
&\leq C \sup_{h \in \mathcal{S}(\mathbb{R}^3);\ \|h\|_{B_{q,1}^{-(\alpha+\beta)}} \leq 1} \int_{\mathbb{R}^3} \frac{\rho_M h}{\nu} f \nu dx \\
&\leq C\|f\|_{B_p^{\alpha+\beta}(\nu)} \sup_{h \in \mathcal{S}(\mathbb{R}^3);\ \|h\|_{B_{q,1}^{-(\alpha+\beta)}} \leq 1} \left\|\frac{\rho_M h}{\nu}\right\|_{B_{q,1}^{-(\alpha+\beta)}(\nu)},
\end{aligned}$$

where C is a constant independent of f and M. Note that $\rho_M/\nu \in C_0^\infty(\mathbb{R}^3)$ and that ρ_M/ν and their derivatives are dominated by CM^σ with a constant C. Since this fact and Proposition 2.4 imply

$$\left\|\frac{\rho_M h}{\nu}\right\|_{B_{q,1}^{-(\alpha+\beta)}(\nu)} \leq CM^\sigma \|h\|_{B_{q,1}^{-(\alpha+\beta)}},$$

we have (2.7). ☐

LEMMA 2.12. *Assume that $\nu(x) = (1+|x|^2)^{-\sigma/2}$ for some $\sigma \in [0,\infty)$. Then, for $M \in \mathbb{N}$ and $p \in [1,\infty]$ there exists a constant C depending on σ, ψ and ρ such that*

$$\left\|P_{M,N} f - \frac{1}{\nu} P_{M,N}(\nu f)\right\|_{L^p(\nu)} \leq CM^\sigma 2^{-N} \|f\|_{L^p(\nu)}$$

for $f \in L^p(\nu)$ and $N \in \mathbb{N}$.

PROOF. It is easy to see that

$$\nu(x)^{1/p}\rho_M(x)\nu(y)^{-1/p}\left(1-\frac{\nu(y)}{\nu(x)}\right)\le\sigma|x-y|\rho_M(x)\nu(x)^{-1}\nu(x-y)^{-1/p}$$

for $x,y\in\mathbb{R}^3$. From this fact, the definition of $P_{M,N}$ and the Plancherel theorem, it follows that

$$\begin{aligned}&\int_{\mathbb{R}^3}\left|P_{M,N}f(x)-\frac{1}{\nu(x)}P_{M,N}(\nu f)(x)\right|^p\nu(x)dx\\&=\int_{\mathbb{R}^3}\left|\int_{\mathbb{R}^3}(\mathcal{F}^{-1}\psi_N)(x-y)\rho_M(x)\left(1-\frac{\nu(y)}{\nu(x)}\right)f(y)dy\right|^p\nu(x)dx\\&\le\sigma\int_{\mathbb{R}^3}\left(\int_{\mathbb{R}^3}|x-y|\nu(x-y)^{-1/p}\left|(\mathcal{F}^{-1}\psi_N)(x-y)\right|\frac{\rho_M(x)}{\nu(x)}|f(y)|\nu(y)^{1/p}dy\right)^p dx.\end{aligned}$$

Hence, by the fact that $\|\rho_M/\nu\|_{L^\infty}\le CM^\sigma$ for each M, and Young's inequality we have

(2.8)

$$\left\|P_{M,N}f-\frac{1}{\nu}P_{M,N}(\nu f)\right\|_{L^p(\nu)}\le CM^\sigma\left(\int_{\mathbb{R}^3}|z|\nu(z)^{-1/p}\left|(\mathcal{F}^{-1}\psi_N)(z)\right|dz\right)\|f\|_{L^p(\nu)}.$$

By the definition of inverse Fourier transform, changing the integration variables and using the integration by parts formula, we get

$$\begin{aligned}&\int_{\mathbb{R}^3}|z|\nu(z)^{-1/p}\left|(\mathcal{F}^{-1}\psi_N)(z)\right|dz\\&=\frac{1}{(2\pi)^{3/2}2^N}\int_{\mathbb{R}^3}|z|(1+2^{-N}|z|)^{\sigma/p}\left|\int_{\mathbb{R}^3}\psi(\xi)e^{\sqrt{-1}z\cdot\xi}d\xi\right|dz.\end{aligned}$$

Then, for any $m\in\mathbb{N}$ the right-hand side of this equality is calculated as

$$\begin{aligned}&\frac{1}{(2\pi)^{3/2}2^N}\int_{\mathbb{R}^3}\frac{|z|(1+2^{-N}|z|)^{\sigma/p}}{(1+|z|^2)^m}\left|\int_{\mathbb{R}^3}\psi(\xi)(1+\triangle_\xi)^m e^{\sqrt{-1}z\cdot\xi}d\xi\right|dz\\&=\frac{1}{(2\pi)^{3/2}2^N}\int_{\mathbb{R}^3}\frac{|z|(1+2^{-N}|z|)^{\sigma/p}}{(1+|z|^2)^m}\left|\int_{\mathbb{R}^3}\left[(1+\triangle)^m\psi\right](\xi)e^{\sqrt{-1}z\cdot\xi}d\xi\right|dz.\end{aligned}$$

But, for $m>2+\sigma/2$ this is dominated by $C2^{-N}$, where C is a constant depending on σ and ψ. From this inequality the proof follows. □

The following propositions are about the estimate of commutators of paraproducts and the heat semigroup, respectively.

PROPOSITION 2.13. Let $\alpha\in(0,1)$, $\beta\in\mathbb{R}$, $\gamma\in(-\infty,0)$ and $p,p_1,p_2,p_3\in[1,\infty]$. Assume that

$$\beta+\gamma<0,\quad\alpha+\beta+\gamma>0,\quad\frac{1}{p}=\frac{1}{p_1}+\frac{1}{p_2}+\frac{1}{p_3}.$$

Then,

$$\begin{aligned}&\left\|\left[\rho(f\varolessthan g)\right]\circleddash h-\rho f(g\circleddash h)\right\|_{B_p^{\alpha+\beta+\gamma}(\nu)}\\&\le C\|\rho\|_{B_\infty^{\alpha+(\beta\vee 0)}(\nu)}\|f\|_{B_{p_1}^\alpha(\nu)}\|g\|_{B_{p_2}^\beta(\nu)}\|h\|_{B_{p_3}^\gamma(\nu)}\end{aligned}$$

for $\rho\in\mathcal{S}(\mathbb{R}^3)$, $f\in B_{p_1}^\alpha(\nu)$, $g\in B_{p_2}^\beta(\nu)$ and $h\in B_{p_3}^\gamma(\nu)$ where C is a constant depending on α, β, γ, p_1, p_2 and p_3.

PROOF. Similarly to [**MW17b**, Proposition A.9] we are able to show that

$$\|(\tilde f \circledless \tilde g)\circleddash \tilde h - \tilde f(\tilde g \circleddash \tilde h)\|_{B_p^{\alpha+\beta+\gamma}(\nu)} \le C\|\tilde f\|_{B_{p_1}^{\alpha}(\nu)}\|\tilde g\|_{B_{p_2}^{\beta}(\nu)}\|\tilde h\|_{B_{p_3}^{\gamma}(\nu)} \tag{2.9}$$

for $\tilde f \in B_{p_1}^{\alpha}(\nu)$, $\tilde g \in B_{p_2}^{\beta}(\nu)$, $\tilde h \in B_{p_3}^{\gamma}(\nu)$ and (α,β,γ) and (p,p_1,p_2,p_3) satisfying the conditions in the statement. By the triangle inequality we have

$$\begin{aligned}
&\| [\rho(f \circledless g)] \circleddash h - \rho f(g \circleddash h)\|_{B_p^{\alpha+\beta+\gamma}(\nu)} \\
&\le \|[\rho \circledless (f \circledless g)] \circleddash h - \rho[(f \circledless g)\circleddash h]\|_{B_p^{\alpha+\beta+\gamma}(\nu)} \\
&\quad + \|\rho\,[(f \circledless g)\circleddash h - f(g \circleddash h)]\|_{B_p^{\alpha+\beta+\gamma}(\nu)} \\
&\quad + \|\,[\rho \circledgtr (f \circledless g)]\circleddash h\|_{B_p^{\alpha+\beta+\gamma}(\nu)}.
\end{aligned} \tag{2.10}$$

Let $q \in [1,\infty]$ such that $1/p = 1/q + 1/p_3$. Applying (2.9) and Proposition 2.5, we have

$$\begin{aligned}
&\|[\rho \circledless (f \circledless g)] \circleddash h - \rho[(f \circledless g)\circleddash h]\|_{B_p^{\alpha+\beta+\gamma}(\nu)} \\
&\le C\,\|\rho\|_{B_\infty^{\alpha}(\nu)}\,\|f \circledless g\|_{B_q^{\beta}(\nu)}\,\|h\|_{B_{p_3}^{\gamma}(\nu)} \\
&\le C\|\rho\|_{B_\infty^{\alpha}(\nu)}\|f\|_{L^{p_1}(\nu)}\|g\|_{B_{p_2}^{\beta}(\nu)}\|h\|_{B_{p_3}^{\gamma}(\nu)}.
\end{aligned} \tag{2.11}$$

Proposition 2.5 and (2.9) imply

$$\begin{aligned}
&\|\rho\,[(f \circledless g)\circleddash h - f(g \circleddash h)]\|_{B_p^{\alpha+\beta+\gamma}(\nu)} \\
&\le C\|\rho\|_{B_\infty^{\alpha+\beta+\gamma}(\nu)}\,\|(f \circledless g)\circleddash h - f(g \circleddash h)\|_{B_p^{\alpha+\beta+\gamma}(\nu)} \\
&\le C\|\rho\|_{B_\infty^{\alpha}(\nu)}\|f\|_{B_{p_1}^{\alpha}(\nu)}\|g\|_{B_{p_2}^{\beta}(\nu)}\|h\|_{B_{p_3}^{\gamma}(\nu)}.
\end{aligned} \tag{2.12}$$

Proposition 2.5 implies

$$\begin{aligned}
&\|\,[\rho \circledgtr (f \circledless g)]\circleddash h\|_{B_p^{\alpha+\beta+\gamma}(\nu)} \\
&\le C\,\|\rho \circledgtr (f \circledless g)\|_{B_q^{\alpha+\beta}(\nu)}\,\|h\|_{B_{p_3}^{\gamma}(\nu)} \\
&\le C\,\|\rho\|_{B_\infty^{\alpha+(\beta\vee 0)}(\nu)}\,\|f \circledless g\|_{B_q^{\beta\wedge 0}(\nu)}\,\|h\|_{B_{p_3}^{\gamma}(\nu)} \\
&\le C\|\rho\|_{B_\infty^{\alpha+(\beta\vee 0)}(\nu)}\|f\|_{B_{p_1}^{\alpha}(\nu)}\|g\|_{B_{p_2}^{\beta}(\nu)}\|h\|_{B_{p_3}^{\gamma}(\nu)}.
\end{aligned} \tag{2.13}$$

Therefore, we have the assertion from (2.10)–(2.13). □

PROPOSITION 2.14. Assume that $\nu(x) = (1+|x|^2)^{-\sigma/2}$ for some $\sigma \in [0,\infty)$. For $\alpha \in [0,1)$, $\beta \in (-\infty,0)$, $\gamma \in \mathbb{R}$ such that $\gamma \ge \alpha+\beta$, $\varepsilon \in (0,\max\{1,1-\alpha\}]$, and $p,p_1,p_2 \in [1,\infty]$ such that $1/p = 1/p_1 + 1/p_2$, there exists a constant C depending on σ, ψ, ρ, α, β, γ and ε such that

$$\begin{aligned}
&\left\|e^{t\triangle}P_N^2\,[\rho_M\,(f \circledless g)] - \rho_M\,\left[f \circledless \left(e^{t\triangle}P_N^2 g\right)\right]\right\|_{B_p^{\gamma}(\nu)} \\
&\le C(1+M^{\sigma}2^{-N})\left(1+t^{-(\gamma-\alpha-\beta)/2}\right)\|f\|_{B_{p_1}^{\alpha+\varepsilon}(\nu)}\|g\|_{B_{p_2}^{\beta}(\nu)}
\end{aligned}$$

for $f,g \in C_0^{\infty}(\mathbb{R}^3)$ such that $\operatorname{supp} f, \operatorname{supp} g \subset \{x \in \mathbb{R}^3; |x| \le 2M\}$, $t \in (0,\infty)$ and $N \in \mathbb{N}$.

PROOF. For simplicity we denote $C\left(1+t^{-(\gamma-\alpha-\beta)/2}\right)$ by C_t, where the constant C can be different from line to line. We remark that similarly to [**AK20**, Proposition 2.6] we are able to show that

$$\left\|e^{t\triangle}P_N^2(\tilde f \circledless \tilde g) - \tilde f \circledless (e^{t\triangle}P_N^2 \tilde g)\right\|_{B_p^{\gamma}(\nu)} \le C_t\|\tilde f\|_{B_{p_1}^{\alpha}(\nu)}\|\tilde g\|_{B_{p_2}^{\beta}(\nu)} \tag{2.14}$$

for $\tilde{f} \in B^{\alpha}_{p_1}(\nu)$ and $\tilde{g} \in B^{\beta}_{p_1}(\nu)$. The triangle inequality and Proposition 2.5 imply

$$\begin{aligned}
&\left\|e^{t\triangle} P_N^2 \left[\rho_M \varolessthan (f \varolessthan g)\right] - \rho_M \varolessthan \left[f \varolessthan \left(e^{t\triangle} P_N^2 g\right)\right]\right\|_{B_p^{\gamma}(\nu)} \\
&\leq \left\|e^{t\triangle} P_N^2 \left[\rho_M \varolessthan (f \varolessthan g)\right] - \rho_M \varolessthan \left[e^{t\triangle} P_N^2 (f \varolessthan g)\right]\right\|_{B_p^{\gamma}(\nu)} \\
&\quad + \left\|\rho_M \varolessthan \left[e^{t\triangle} P_N^2 (f \varolessthan g) - f \varolessthan \left(e^{t\triangle} P_N^2 g\right)\right]\right\|_{B_p^{\gamma}(\nu)} \\
&\leq \left\|e^{t\triangle} P_N^2 \left[\rho_M \varolessthan (f \varolessthan g)\right] - \rho_M \varolessthan \left[e^{t\triangle} P_N^2 (f \varolessthan g)\right]\right\|_{B_p^{\gamma}(\nu)} \\
&\quad + \|\rho_M\|_{L^{\infty}(\nu)} \left\|e^{t\triangle} P_N^2 (f \varolessthan g) - f \varolessthan \left(e^{t\triangle} P_N^2 g\right)\right\|_{B_p^{\gamma}(\nu)}.
\end{aligned}$$

Hence, applying (2.14) and Proposition 2.5, we have

$$\begin{aligned}
&\left\|e^{t\triangle} P_N^2 \left[\rho_M \varolessthan (f \varolessthan g)\right] - \rho_M \varolessthan \left[f \varolessthan \left(e^{t\triangle} P_N^2 g\right)\right]\right\|_{B_p^{\gamma}(\nu)} \\
&\leq C_t \|\rho_M\|_{B^{\alpha}_{\infty}(\nu)} \|f \varolessthan g\|_{B^{\beta}_{p}(\nu)} + C_t \|f\|_{B^{\alpha}_{p_1}(\nu)} \|g\|_{B^{\beta}_{p_2}(\nu)} \\
&\leq C_t \left(\|f\|_{L^{p_1}(\nu)} \|g\|_{B^{\beta}_{p_2}(\nu)} + \|f\|_{B^{\alpha}_{p_1}(\nu)} \|g\|_{B^{\beta}_{p_2}(\nu)}\right).
\end{aligned}$$

In view of this inequality, to finish the proof it is sufficient to show that

$$\left\|e^{t\triangle} P_N^2 \left[\rho_M \varogreaterthan (f \varolessthan g)\right]\right\|_{B_p^{\gamma}(\nu)} \leq C_t M^{\sigma} \|f\|_{B^{\alpha+\varepsilon}_{p_1}(\nu)} \|g\|_{B^{\beta}_{p_2}(\nu)} \tag{2.15}$$

$$\left\|\rho_M \varogreaterthan \left[f \varolessthan \left(e^{t\triangle} P_N^2 g\right)\right]\right\|_{B_p^{\gamma}(\nu)} \leq C M^{\sigma} \|f\|_{B^{\alpha}_{p_1}(\nu)} \|g\|_{B^{\beta}_{p_2}(\nu)}. \tag{2.16}$$

[**BCD11**, Proposition 2.76], Proposition 2.4 and the support of g imply

$$\begin{aligned}
\|g\|_{B^{\beta}_{p_2}} &\leq C \sup_{h \in \mathcal{S}(\mathbb{R}^3);\ \|h\|_{B^{-\beta}_{p_1',1}} \leq 1} \langle h, g\rangle \\
&\leq C \sup_{h \in \mathcal{S}(\mathbb{R}^3);\ \|h\|_{B^{-\beta}_{p_1',1}} \leq 1} \int_{\mathbb{R}^3} \frac{\rho_{2M} h}{\nu} g \nu dx \\
&\leq C \|g\|_{B^{\beta}_{p_2}(\nu)} \sup_{h \in \mathcal{S}(\mathbb{R}^3);\ \|h\|_{B^{-\beta}_{p_1',1}} \leq 1} \left\|\frac{\rho_{2M} h}{\nu}\right\|_{B^{-\beta}_{p_1'}(\nu)}.
\end{aligned}$$

Note that $\rho_{2M}/\nu \in C_0^{\infty}(\mathbb{R}^3)$ and that ρ_M/ν and their derivatives are dominated by CM^{σ} with a constant C. Since this fact implies

$$\left\|\frac{\rho_{2M} h}{\nu}\right\|_{B^{-\beta}_{p_1',1}(\nu)} \leq C M^{\sigma} \|h\|_{B^{-\beta}_{p_1',1}},$$

we have

$$\|g\|_{B^{\beta}_{p_2}} \leq C M^{\sigma} \|g\|_{B^{\beta}_{p_2}(\nu)}. \tag{2.17}$$

By the triangle inequality, Propositions 2.4 and 2.5, Lemma 2.10 and the supports of f and g, we have

$$
\begin{aligned}
&\left\|e^{t\triangle} P_N^2 \left[\rho_M \circledcirc (f \circledcirc g)\right]\right\|_{B_p^\gamma(\nu)} \\
&\le \left\|e^{t\triangle} \left[\rho_M \circledcirc (f \circledcirc g)\right]\right\|_{B_p^\gamma(\nu)} + \left\|(I - P_N^2) e^{t\triangle} \left[\rho_M \circledcirc (f \circledcirc g)\right]\right\|_{B_p^\gamma(\nu)} \\
&\le C_t \left\|\rho_M \circledcirc (f \circledcirc g)\right\|_{B_p^{\alpha+\beta}(\nu)} + \left\|(I - P_N^2) e^{t\triangle} \left[\rho_M \circledcirc (f \circledcirc g)\right]\right\|_{B_p^\gamma} \\
&\le C_t \left\|\rho_M\right\|_{B_\infty^{\alpha+1}(\nu)} \left\|f \circledcirc g\right\|_{B_p^\beta(\nu)} + C_t 2^{-N} \left\|\rho_M \circledcirc (f \circledcirc g)\right\|_{B_p^{\alpha+\beta+1}} \\
&\le C_t \left(\left\|f \circledcirc g\right\|_{B_p^\beta(\nu)} + 2^{-N} \left\|\rho_M\right\|_{B_\infty^{\alpha+2}} \left\|f \circledcirc g\right\|_{B_p^\beta}\right) \\
&\le C_t \left(\|f\|_{L^{p_1}(\nu)} \|g\|_{B_{p_2}^\beta(\nu)} + 2^{-N} \|f\|_{L^{p_1}(1)} \|g\|_{B_{p_2}^\beta}\right) \\
&\le C_t \left(\|f\|_{B_{p_1}^{\alpha+\varepsilon}(\nu)} \|g\|_{B_{p_2}^\beta(\nu)} + C 2^{-N} \|f\|_{L^{p_1}(\nu)} \|g\|_{B_{p_2}^\beta}\right).
\end{aligned}
$$

Thus, (2.15) follows from (2.17). From Propositions 2.4 and 2.5, we have

$$
\begin{aligned}
&\left\|\rho_M \circledcirc \left[f \circledcirc \left(e^{t\triangle} P_N^2 g\right)\right]\right\|_{B_p^\gamma(\nu)} \\
&\le C \left\|\rho_M\right\|_{B_\infty^{-\beta+|\gamma|+2}(\nu)} \left\|f \circledcirc \left(e^{t\triangle} P_N^2 g\right)\right\|_{B_p^{\beta-1}(\nu)} \\
&\le C \|f\|_{L^{p_1}(\nu)} \left\|e^{t\triangle} P_N^2 g\right\|_{B_{p_2}^{\beta-1}(\nu)} \\
&\le C \|f\|_{B_{p_1}^\varepsilon(\nu)} \left\|P_N^2 g\right\|_{B_{p_2}^{\beta-1}(\nu)} \\
&\le C \|f\|_{B_{p_1}^\varepsilon(\nu)} \left(\|g\|_{B_{p_2}^{\beta-1}(\nu)} + \left\|(I - P_N^2) g\right\|_{B_{p_2}^{\beta-1}(\nu)}\right).
\end{aligned}
$$

Since Lemma 2.10 and (2.17) imply

$$
\left\|(I - P_N^2) g\right\|_{B_{p_2}^{\beta-1}(\nu)} \le \left\|(I - P_N^2) g\right\|_{B_{p_2}^{\beta-1}} \le C 2^{-N} \|g\|_{B_{p_2}^\beta} \le C M^\sigma 2^{-N} \|g\|_{B_{p_2}^\beta(\nu)},
$$

(2.16) follows. □

CHAPTER 3

Infinite-dimensional Ornstein-Uhlenbeck process

In this section we introduce the relevant infinite-dimensional Ornstein-Uhlenbeck process and its polynomials. The polynomials of the Ornstein-Uhlenbeck process appear in the renormalization and the transformation of the stochastic partial differential equation associated to the Φ^4_3-model on the full Euclidean space $\mathbb{R}^3$.

Let μ_0 be the centered Gaussian measure on $\mathcal{S}'(\mathbb{R}^3)$ with the covariance operator $[2(-\triangle + m_0^2)]^{-1}$ where $m_0 > 0$ as in Section 1.2, let $\dot{W}_t(x)$ be a Gaussian white noise with parameter $(t, x) \in (-\infty, \infty) \times \mathbb{R}^3$, and let W_t be the cylindrical Brownian motion on $L^2(dx)$ given by

$$\langle f, W_t \rangle = \begin{cases} {}_{\mathcal{S}((-\infty,\infty)\times\mathbb{R}^3)}\langle f \otimes \mathbb{I}_{[0,t]}, \dot{W} \rangle_{\mathcal{S}'((-\infty,\infty)\times\mathbb{R}^3)}, & t \in [0, \infty) \\ {}_{\mathcal{S}((-\infty,\infty)\times\mathbb{R}^3)}\langle f \otimes \mathbb{I}_{[t,0]}, \dot{W} \rangle_{\mathcal{S}'((-\infty,\infty)\times\mathbb{R}^3)}, & t \in (-\infty, 0), \end{cases}$$

for $f \in \mathcal{S}(\mathbb{R}^3)$. We define the $\mathcal{S}'(\mathbb{R}^3)$-valued stochastic process Z_t by the stochastic convolution

$$Z_t = \int_{-\infty}^{t} e^{(t-s)(\triangle - m_0^2)} dW_s, \quad t \in (-\infty, \infty). \tag{3.1}$$

We remark that (3.1) is an equation not only for positive t, but also for negative t. It is well-known that Z_t is the infinite-dimensional Ornstein-Uhlenbeck process and the law of Z_t is μ_0 for $t \in (-\infty, \infty)$. Hence, Z_t is the stationary Ornstein-Uhlenbeck process. Moreover, Z_t is the mild solution to the stochastic partial differential equation on $\mathbb{R}^3$:

$$\partial_t Z_t(x) = \dot{W}_t(x) - (-\triangle + m_0^2) Z_t(x), \quad (t, x) \in (-\infty, \infty) \times \mathbb{R}^3. \tag{3.2}$$

To find the covariance operator of our approximation $P_N Z_t$ of Z_t with P_N as in Section 1.2, we prepare a lemma as follows.

LEMMA 3.1. *For $f, g \in \mathcal{S}(\mathbb{R}^3)$ and $s, t \in (-\infty, \infty)$ such that $s < t$*

$$E\left[\langle f, P_N Z_t \rangle \langle g, P_N Z_s \rangle\right] = \langle V_N(t - s) f, g \rangle$$

where $\{V_N(t) : t \in [0, \infty)\}$ is a family of bounded linear operators on $B^{\alpha}_{p,r}$ for $\alpha \in \mathbb{R}$ and $p, r \in [1, \infty]$ given by

$$V_N(t) f := P_N^2 [2(m_0^2 - \triangle)]^{-1} e^{t(\triangle - m_0^2)} f.$$

Moreover, the following estimates hold:

(i) *For $\alpha \in \mathbb{R}$ and $\beta, \gamma \in [0, \infty)$*

$$\|V_N(t) f\|_{B^{\alpha+\beta+\gamma}_{p,r}} \leq C 2^{\beta N} (1 + t^{-\gamma/2}) e^{-m_0^2 t} \|f\|_{B^{\alpha-2}_{p,r}}, \quad t \in (0, \infty), \ f \in \mathcal{S}(\mathbb{R}^3),$$

where C is a constant depending on α, β and γ.

(ii) *For* $\alpha \in \mathbb{R}$ *and* $\beta \in [0, \infty)$

$$\|V_N(0)f\|_{B^{\alpha+\beta}_{p,r}} \leq C2^{\beta N}\|f\|_{B^{\alpha-2}_{p,r}}, \quad f \in \mathcal{S}(\mathbb{R}^3)$$

where C *is a constant depending on* α *and* β.

(iii) *For* $\alpha \in \mathbb{R}$, $\beta \in [0, \infty)$ *and* $\varepsilon \in (0, 1)$

$$\sum_{N=1}^{\infty} 2^{-\beta N} \left\|(P_{N+1} - P_N)\left[2(m_0^2 - \triangle)\right]^{-1} f\right\|_{B^{\alpha+\beta-\varepsilon}_{p,r}} \leq C\|f\|_{B^{\alpha-2}_{p,r}}$$

for $f \in \mathcal{S}(\mathbb{R}^3)$, *where* C *is a constant depending on* α, β *and* ε.

(iv) *For* $\alpha \in \mathbb{R}$, $\beta, \gamma \in [0, \infty)$ *and* $s, t \in [0, \infty)$ *such that* $s < t$

$$\|(V_N(t) - V_N(s))f\|_{B^{\alpha+\beta}_{p,r}} \leq C2^{\beta N}(t-s)^{\gamma/2}\|f\|_{B^{\alpha+\gamma-2}_{p,r}}, \quad f \in \mathcal{S}(\mathbb{R}^3),$$

where C *is a constant depending on* α, β *and* γ.

PROOF. For fixed s, t, consider the time reversal $\tilde{W}_u := W_{s-u}$. Then, we have

$$\begin{aligned}
&E\left[\langle f, P_N Z_t\rangle\langle g, P_N Z_s\rangle\right] \\
&= E\left[\int_{-\infty}^{t} \left\langle e^{(t-u)(\triangle - m_0^2)} P_N f, dW_u\right\rangle \int_{-\infty}^{s} \left\langle e^{(s-u)(\triangle - m_0^2)} P_N g, dW_u\right\rangle\right] \\
&= E\left[\int_{s-t}^{\infty} \left\langle e^{(t-s+u)(\triangle - m_0^2)} P_N f, d\tilde{W}_u\right\rangle \int_{0}^{\infty} \left\langle e^{u(\triangle - m_0^2)} P_N g, d\tilde{W}_u\right\rangle\right].
\end{aligned}$$

Hence, by the fact that $\tilde{W}_u$ is also the white noise, Itô's isometry yields

$$\begin{aligned}
&E\left[\langle f, P_N Z_t\rangle\langle g, P_N Z_s\rangle\right] \\
&= \int_0^{\infty} \left\langle e^{(t-s+u)(\triangle - m_0^2)} P_N f, e^{u(\triangle - m_0^2)} P_N g\right\rangle du \\
&= \left\langle P_N \left(\int_0^{\infty} e^{2u(\triangle - m_0^2)} du\right) e^{(t-s)(\triangle - m_0^2)} P_N f, g\right\rangle.
\end{aligned}$$

In view of the commutativity between P_N and the heat semigroup, we have the first assertion. To show (i), note that $\int_0^\infty e^{2u(\triangle - m_0^2)} du$ coincides with the resolvent operator $[2(m_0^2 - \triangle)]^{-1}$, and $[2(m_0^2 - \triangle)]^{-1}$ is a bounded operator from $B^{\alpha-2}_{p,r}$ to $B^{\alpha}_{p,r}$. By using this fact, Proposition 2.5 and Lemmas 2.9 and 2.10, we calculate that

$$\begin{aligned}
\|V_N(t)f\|_{B^{\alpha+\beta+\gamma}_{p,r}} &\leq C\left\|P_N^2[2(m_0^2 - \triangle)]^{-1} e^{t(\triangle - m_0^2)} f\right\|_{B^{\alpha+\beta+\gamma}_{p,r}} \\
&\leq C2^{\beta N}(1 + t^{-\gamma/2})e^{-m_0^2 t}\left\|P_N^2 f\right\|_{B^{\alpha-2}_{p,r}} \\
&\leq C2^{\beta N}(1 + t^{-\gamma/2})e^{-m_0^2 t}\|f\|_{B^{\alpha-2}_{p,r}}.
\end{aligned}$$

Hence, the estimate for $t \in (0, \infty)$ holds. Similarly, we have (ii). We show (iii) in the case that $r \in [1, \infty)$. Let $g = [2(m_0^2 - \triangle)]^{-1} f$. In view of the support of φ and

ψ, we have

$$\begin{aligned}
&\sum_{N=1}^{\infty} 2^{-\beta N} \|(P_{N+1} - P_N)g\|_{B^{\alpha+\beta-\varepsilon}_{p,r}} \\
&= \sum_{N=1}^{\infty} 2^{-\beta N} \left(\sum_{j=-1}^{\infty} 2^{(\alpha+\beta-\varepsilon)rj} \|(P_{N+1} - P_N)\Delta_j g\|_{L^p}^r \right)^{1/r} \\
&\leq C \sum_{N=1}^{\infty} 2^{-\varepsilon N} \left(\sum_{j=-1}^{\infty} 2^{\alpha rj} \|(P_{N+1} - P_N)^2 \Delta_j g\|_{L^p}^r \right)^{1/r}.
\end{aligned}$$

Since Young's inequality and a equation of ψ_N similar to (2.5) imply

$$\begin{aligned}
\|(P_{N+1} - P_N)\Delta_j g\|_{L^p} &= \left|(\mathcal{F}^{-1}\psi_{N+1} - \mathcal{F}^{-1}\psi_N) * \Delta_j g\right\|_{L^p} \\
&\leq \left\|\mathcal{F}^{-1}\psi_{N+1} - \mathcal{F}^{-1}\psi_N\right\|_{L^1} \|\Delta_j g\|_{L^p},
\end{aligned}$$

and

$$\begin{aligned}
\left\|\mathcal{F}^{-1}\psi_{N+1} - \mathcal{F}^{-1}\psi_N\right\|_{L^1} &= 2^{3N} \left\|\left(\mathcal{F}^{-1}\psi_1\right)(2^N \cdot) - \left(\mathcal{F}^{-1}\psi\right)(2^N \cdot)\right\|_{L^1} \\
&= \left\|\mathcal{F}^{-1}\psi_1 - \mathcal{F}^{-1}\psi\right\|_{L^1},
\end{aligned}$$

respectively, we have

$$\sum_{N=1}^{\infty} 2^{-\beta N} \|(P_{N+1} - P_N)g\|_{B^{\alpha+\beta-\varepsilon}_{p,r}} \leq C \|g\|_{B^{\alpha}_{p,r}}.$$

Hence, (iii) follows from the fact that $\|g\|_{B^{\alpha}_{p,r}} \leq C \|f\|_{B^{\alpha-2}_{p,r}}$. The case where $r = \infty$ is similar. So we omit it.

We also obtain (iv) similarly, because

$$\left\|\left(e^{t(\triangle - m_0^2)} - e^{s(\triangle - m_0^2)}\right) f\right\|_{B^{\alpha-2}_{p,r}} \leq C(t-s)^{\gamma/2} \|f\|_{B^{\alpha+\gamma-2}_{p,r}}. \qquad \square$$

Now we consider the renormalization of polynomials of the Ornstein-Uhlenbeck processes. Let

$$\begin{aligned}
C_1^{(N)} &:= \langle V_N(0)\delta_0, \delta_0 \rangle \\
C_2^{(M,N)}(x) &:= 2 \sum_{i,j=-1}^{\infty} \mathbb{I}_{[-1,1]}(i-j) \\
&\quad \times \int_0^{\infty} e^{t(\triangle - m_0^2)} P_N^2 \left[\left(\Delta_i P_N \left[\rho_M^2 V_N(t) \left(\rho_M^2 \Delta_j P_N \delta_x \right) \right] \right)^2 \right](x) dt
\end{aligned}$$

where δ_x is the Dirac delta function. Note that $C_1^{(N)}$ and $C_2^{(M,N)}(x)$ are same as those in Section 1.2. Define

$$\begin{aligned}
\mathcal{Z}_t^{(1,N)} &:= P_N Z_t,\\
\mathcal{Z}_t^{(2,N)} &:= (P_N Z_t)^2 - C_1^{(N)},\\
\mathcal{Z}_t^{(3,N)} &:= (P_N Z_t)^3 - 3C_1^{(N)} P_N Z_t,\\
\mathcal{Z}_t^{(1,M,N)} &:= P_{M,N} Z_t,\\
\mathcal{Z}_t^{(2,M,N)} &:= (P_{M,N} Z_t)^2 - \rho_M^2 C_1^{(N)},\\
\mathcal{Z}_t^{(3,M,N)} &:= (P_{M,N} Z_t)^3 - 3\rho_M^2 C_1^{(N)} P_{M,N} Z_t,\\
\mathcal{Z}_t^{(0,2,M,N)} &:= \int_{-\infty}^t e^{(t-s)(\triangle - m_0^2)} P_{M,N}^* \mathcal{Z}_s^{(2,M,N)} ds,\\
\mathcal{Z}_t^{(0,3,M,N)} &:= \int_{-\infty}^t e^{(t-s)(\triangle - m_0^2)} P_{M,N}^* \mathcal{Z}_s^{(3,M,N)} ds,\\
\mathcal{Z}_t^{(2,2,M,N)} &:= \mathcal{Z}_t^{(2,M,N)} \circleddash \int_{-\infty}^t e^{(t-s)(\triangle - m_0^2)} P_N^2 \mathcal{Z}_s^{(2,M,N)} ds - C_2^{(M,N)},\\
\mathcal{Z}_t^{(2,3,M,N)} &:= \mathcal{Z}_t^{(2,M,N)} \circleddash \int_{-\infty}^t e^{(t-s)(\triangle - m_0^2)} P_N^2 \mathcal{Z}_s^{(3,M,N)} ds - 3C_2^{(M,N)} \mathcal{Z}_t^{(1,M,N)},\\
\widehat{\mathcal{Z}}_t^{(2,3,M,N)} &:= \mathcal{Z}_t^{(2,M,N)} \circleddash \left(P_{M,N} \mathcal{Z}_t^{(0,3,M,N)}\right) - 3C_2^{(M,N)} \rho_M^2 \mathcal{Z}_t^{(1,M,N)}
\end{aligned}$$

for $t \in (-\infty, \infty)$ and $N \in \mathbb{N}$. We remark that

$$\mathcal{Z}_t^{(k,M,N)} = \rho_M^k \mathcal{Z}_t^{(k,N)}$$

for $k = 1, 2, 3$. Moreover, in view of the stationarity of Z, by explicit calculation (see the proof of Proposition 3.3) we have

$$\begin{aligned}
C_1^{(N)} &= E\left[(P_N Z_t)^2(x)\right]\\
C_2^{(M,N)}(x) &= E\left[\left(\mathcal{Z}_t^{(2,M,N)} \circleddash \int_{-\infty}^t e^{(t-s)(\triangle - m_0^2)} P_N^2 \mathcal{Z}_s^{(2,M,N)} ds\right)(x)\right]
\end{aligned}$$

for $x \in \mathbb{R}^3$ and $M, N \in \mathbb{N}$. In particular, $C_1^{(N)}$ is independent of x and t, and $\mathcal{Z}^{(k,N)}$ is a Wick polynomial of $P_N Z$ for $k = 1, 2, 3$. For later use, we see asymptotics of $C_1^{(N)}$ and $C_2^{(M,N)}(x)$ as N goes to infinity.

PROPOSITION 3.2. There exist positive constants c_1 and c_2 such that

$$\begin{aligned}
&\lim_{N\to\infty} 2^{-N} C_1^{(N)} = c_1,\\
&-c_2 \le C_2^{(M,N)}(x) \le c_2 N = \frac{c_2}{\log 2} \cdot \log 2^N
\end{aligned}$$

for $x \in \mathbb{R}^3$ and $M, N \in \mathbb{N}$.

PROOF. Since

$$\begin{aligned}
\left[P_N^2 \delta_0\right](x) &= (2\pi)^{-3/2} \left[\mathcal{F}^{-1}(\psi_N^2)\right](x)\\
&= (2\pi)^{-3/2} 2^{3N} \left[\mathcal{F}^{-1}(\psi^2)\right](2^N x),\\
\left[[2(m_0^2 - \triangle)]^{-1}\delta_0\right](x) &= \int_0^\infty (8\pi u)^{-3/2} \exp\left(-\frac{|x|^2}{8u} - 2m_0^2 u\right) du,
\end{aligned}$$

we are able to calculate $C_1^{(N)}$ explicitly as

$$\begin{aligned}
C_1^{(N)} &= \frac{2^{3N}}{8(2\pi)^3}\int_0^\infty \int_{\mathbb{R}^3} u^{-3/2}\left[\mathcal{F}^{-1}(\psi^2)\right](2^N x)\exp\left(-\frac{|x|^2}{8u} - 2m_0^2 u\right) dxdu \\
&= \frac{2^N}{8(2\pi)^3}\int_0^\infty \int_{\mathbb{R}^3} v^{-3/2}\left[\mathcal{F}^{-1}(\psi^2)\right](y)\exp\left(-\frac{|y|^2}{8v} - 2^{-2N+1}m_0^2 v\right) dydv \\
&= \frac{2^N}{(2\pi)^{3/2}}\int_0^\infty \int_{\mathbb{R}^3} (\psi^2)(\xi)\exp\left(-\frac{4v|\xi|^2}{2} - 2^{-2N+1}m_0^2 v\right) d\xi dv \\
&= \frac{2^N}{(2\pi)^{3/2}}\int_{\mathbb{R}^3} \frac{(\psi^2)(\xi)}{2|\xi|^2 + 2^{-2N+1}m_0^2} d\xi.
\end{aligned}$$

Hence, by the compactness of the support of ψ we have the first assertion.

For $i \in \mathbb{N}\cup\{-1,0\}$, $\alpha \in \mathbb{R}$ and $x \in \mathbb{R}^3$, the fact that $\Delta_i\Delta_j = 0$ for $|i-j| \geq 2$ and the Plancherel theorem imply

$$\begin{aligned}
\|\Delta_i \delta_x\|^2_{B^\alpha_{2,2}} &= \sum_{j=-1}^\infty 2^{2\alpha j}\|\Delta_i\Delta_j\delta_x\|^2_{L^2} \\
&\leq C2^{2\alpha i}\|\Delta_i\delta_x\|^2_{L^2} \\
&= C2^{2\alpha i}\|\psi_i \mathcal{F}\delta_x\|^2_{L^2}.
\end{aligned}$$

Hence, by the explicit calculation:

$$\|\psi_i\mathcal{F}\delta_x\|^2_{L^2} = \frac{1}{(2\pi)^{3/2}}\|\psi_i\|^2_{L^2} = \frac{2^{3i}}{(2\pi)^{3/2}}\|\psi\|^2_{L^2},$$

we have

$$\|\Delta_i\delta_x\|^2_{B^\alpha_{2,2}} \leq C2^{(2\alpha+3)i} \tag{3.3}$$

for $i \in \mathbb{N}\cup\{-1,0\}$, $\alpha \in \mathbb{R}$ and $x \in \mathbb{R}^3$. On the other hand, Proposition 2.5 and Lemma 3.1 imply that

$$\begin{aligned}
&\left\|\Delta_i P_N\left[\rho_M^2 V_N(t)\left(\rho_M^2\Delta_j P_N\delta_x\right)\right]\right\|_{B^{3/2}_{2,2}} \\
&\leq C\|\rho_M^2\|_{B^{1/2+\beta}_{2,2}}\left(\left\|[2(m_0^2-\triangle)]^{-1}e^{t(\triangle-m_0^2)}P_N^2\left(\rho_M^2 \circledcirc \Delta_j P_N\delta_x\right)\right\|_{B^{3/2}_{2,2}}\right. \\
&\qquad\qquad\left. + \left\|V_N(t)\left(\rho_M^2 \circledcirc \Delta_j P_N\delta_x\right)\right\|_{B^{3/2}_{2,2}}\right) \\
&\leq C\left(\left\|e^{t(\triangle-m_0^2)}P_N^2\left(\rho_M^2 \circledcirc \Delta_j P_N\delta_x\right)\right\|_{B^{-1/2}_{2,2}} + \left\|\rho_M^2 \circledcirc \Delta_j P_N\delta_x\right\|_{B^{-1/2}_{2,2}}\right).
\end{aligned}$$

Since Proposition 2.4 and (2.14) imply that for sufficiently small $\varepsilon \in (0,1]$

$$\begin{aligned}
&\left\|e^{t(\triangle-m_0^2)}P_N^2\left(\rho_M^2 \circledcirc \Delta_j P_N\delta_x\right)\right\|_{B^{-1/2}_{2,2}} \\
&\leq \left\|e^{t(\triangle-m_0^2)}P_N^2\left(\rho_M^2 \circledcirc \Delta_j P_N\delta_x\right) - \rho_M^2 \circledcirc \left(e^{t(\triangle-m_0^2)}P_N^2\Delta_j P_N\delta_x\right)\right\|_{B^{-1/2+\varepsilon}_{2,\infty}} \\
&\qquad + \left\|\rho_M^2 \circledcirc \left(e^{t(\triangle-m_0^2)}P_N^2(\Delta_j P_N\delta_x)\right)\right\|_{B^{-1/2}_{2,2}} \\
&\leq C(1+t^{-3\varepsilon/2})\|\rho_M^2\|_{B^{1-\varepsilon}_{\infty,\infty}}\|\Delta_j P_N\delta_x\|_{B^{-3/2-\varepsilon}_{2,\infty}} \\
&\qquad + C\|\rho_M^2\|_{L^\infty}\left\|P_N^3(e^{t(\triangle-m_0^2)}\Delta_j\delta_x)\right\|_{B^{-1/2}_{2,2}},
\end{aligned}$$

by Propositions 2.4 and 2.5, Lemma 2.10, [**BCD11**, Lemma 2.4] and (3.3) we have

$$
\begin{aligned}
&\left\|\Delta_i P_N\left[\rho_M^2 V_N(t)\left(\rho_M^2 \Delta_j P_N \delta_x\right)\right]\right\|_{B_{2,2}^{3/2}}^2 \\
&\leq C\Bigg(\left(1+t^{-3\varepsilon/2}\right)\left\|\Delta_j P_N \delta_x\right\|_{B_{2,2}^{-3/2}}+\left\|P_N^3\left(e^{t(\triangle-m_0^2)}\Delta_j \delta_x\right)\right\|_{B_{2,2}^{-1/2}} \\
&\qquad\qquad +\left\|\rho_M^2 \circledcirc \Delta_j P_N \delta_x\right\|_{B_{2,2}^{-3/2}}\Bigg)^2 \\
&\leq C\Bigg(\left(1+t^{-3\varepsilon/2}\right)\left\|\Delta_j \delta_x\right\|_{B_{2,2}^{-3/2}}+e^{-t2^{2j+1}}\left\|\Delta_j \delta_x\right\|_{B_{2,2}^{-1/2}} \\
&\qquad\qquad +\left\|\rho_M^2\right\|_{B_{\infty,\infty}^2}\left\|\Delta_j \delta_x\right\|_{B_{2,2}^{-3/2}}\Bigg)^2 \\
&\leq C\left(1+t^{-3\varepsilon}+2^{2j} e^{-t2^{2j+1}}\right).
\end{aligned}
$$

From this inequality, Propositions 2.4 and 2.5, Lemma 2.10 and Corollary 2.3, we have

$$
\begin{aligned}
&\left\|e^{t(\triangle-m_0^2)} P_N^2\left[\left(\Delta_i P_N\left[\rho_M^2 V_N(t)\left(\rho_M^2 \Delta_j P_N \delta_x\right)\right]\right)^2\right]\right\|_{L^\infty} \\
&\leq C\left\|e^{t(\triangle-m_0^2)} P_N^2\left[\left(\Delta_i P_N\left[\rho_M^2 V_N(t)\left(\rho_M^2 \Delta_j P_N \delta_x\right)\right]\right)^2\right]\right\|_{B_{\infty,1}^0} \\
&\leq C e^{-m_0^2 t}\left\|\left(\Delta_i P_N\left[\rho_M^2 V_N(t)\left(\rho_M^2 \Delta_j P_N \delta_x\right)\right]\right)^2\right\|_{B_{1,1}^{3/2}} \\
&\leq C e^{-m_0^2 t}\left\|\Delta_i P_N\left[\rho_M^2 V_N(t)\left(\rho_M^2 \Delta_j P_N \delta_x\right)\right]\right\|_{B_{2,2}^{3/2}}^2 \\
&\leq C e^{-m_0^2 t}\left(1+t^{-3\varepsilon}+2^{2j} e^{-t2^{2j+1}}\right).
\end{aligned}
$$

In view of this inequality and (2.6), there exists $j_0 \in \mathbb{N}$ independent of M and N such that

$$
\begin{aligned}
&\left\|\sum_{i,j=-1}^{\infty} \mathbb{I}_{[-1,1]}(i-j) \int_0^\infty e^{t(\triangle-m_0^2)} P_N^2\left[\left(\Delta_i P_N\left[\rho_M^2 V_N(t)\left(\rho_M^2 \Delta_j P_N \delta_x\right)\right]\right)^2\right] dt\right\|_{L^\infty} \\
&\leq C \sum_{i,j=-1}^{N+j_0} \mathbb{I}_{[-1,1]}(i-j) \int_0^\infty e^{-m_0^2 t}\left(1+t^{-3\varepsilon}+2^{2j} e^{-t2^{2j+1}}\right) dt \\
&\leq CN.
\end{aligned}
$$

Thus, we have the upper bound of $C_2^{M,N}(x)$.

Since for $f \in \mathcal{S}'(\mathbb{R}^3)$ the Fourier transform of $(\Delta_i f)^2$ is supported in $\{x \in \mathbb{R}^3; |x| \leq (16/3)2^i\}$, there exists $j_0 \in \mathbb{N}$ independent of M and N such that

$$
\begin{aligned}
&C_2^{(M,N)}(x) \\
&= 2 \sum_{i,j=-1}^{N-j_0-1} \mathbb{I}_{[-1,1]}(i-j) \int_0^\infty e^{t(\triangle - m_0^2)} \left[\left(\Delta_i P_N \left[\rho_M^2 V_N(t) \left(\rho_M^2 \Delta_j P_N \delta_x \right) \right] \right)^2 \right](x) dt \\
&\quad + 2 \sum_{i,j=N-j_0}^{N+j_0} \mathbb{I}_{[-1,1]}(i-j) \\
&\qquad \times \int_0^\infty e^{t(\triangle - m_0^2)} P_N^2 \left[\left(\Delta_i P_N \left[\rho_M^2 V_N(t) \left(\rho_M^2 \Delta_j P_N \delta_x \right) \right] \right)^2 \right](x) dt \\
&\geq -C \sup_{i,j \in \mathbb{N} \cup \{-1,0\}} \int_0^\infty \left\| e^{t(\triangle - m_0^2)} P_N^2 \left[\left(\Delta_i P_N \left[\rho_M^2 V_N(t) \left(\rho_M^2 \Delta_j P_N \delta_x \right) \right] \right)^2 \right] \right\|_{L^\infty} dt.
\end{aligned}
$$

Hence, by the estimate above we have the lower bound of $C_2^{M,N}(x)$. □

Now we discuss the functionals of $P_N Z$ defined above. Similarly to [**AK20**, Proposition 3.3], the following proposition holds. However, we are not able to use the Fourier expansion now, because we consider the Φ^4-model not on a torus as it was the case in [**AK20**], but on $\mathbb{R}^3$. Instead of the Fourier expansion, we apply estimates of the covariance operator that appeared in Lemma 3.1 above.

PROPOSITION 3.3. *Let $\varepsilon \in (0,1]$ and $\nu \in L^1(dx)$ such that $0 \leq \nu \leq 1$. Then for all $T \in (0,\infty)$, $p \in [1,\infty)$ and $\alpha \in [0,\infty)$ the following properties hold:*

(i) *For $k = 1,2$, $\{2^{-\alpha N} \mathcal{Z}^{(k,N)}; N \in \mathbb{N}\}$ converges almost surely in $C([0,\infty); B_{p,p}^{\alpha-(k/2)-\varepsilon}(\nu))$ and satisfies*

$$
\sup_{N \in \mathbb{N}} 2^{-p\alpha N} E\left[\sup_{t \in [0,T]} \left\| \mathcal{Z}_t^{(k,N)} \right\|_{B_{p,p}^{\alpha-(k/2)-\varepsilon}(\nu)}^p \right] < \infty.
$$

In particular, $\{2^{-\alpha N} \mathcal{Z}^{(k,M,N)}; N \in \mathbb{N}\}$ converges almost surely in $C([0,\infty); B_{p,p}^{\alpha-(k/2)-\varepsilon}(\nu))$ for each M and satisfies

$$
\sup_{M,N \in \mathbb{N}} 2^{-p\alpha N} E\left[\sup_{t \in [0,T]} \left\| \mathcal{Z}_t^{(k,M,N)} \right\|_{B_{p,p}^{\alpha-(k/2)-\varepsilon}(\nu)}^p \right] < \infty.
$$

(ii) *For $k = 2,3$, $\{2^{-\alpha N} \mathcal{Z}^{(0,k,M,N)}; N \in \mathbb{N}\}$ converges almost surely in $C([0,\infty); B_{p,p}^{\alpha+2-(k/2)-\varepsilon}(\nu))$ for each M and satisfies*

$$
\sup_{M,N \in \mathbb{N}} 2^{-p\alpha N} E\left[\sup_{t \in [0,T]} \left\| \mathcal{Z}_t^{(0,k,M,N)} \right\|_{B_{p,p}^{\alpha+2-(k/2)-\varepsilon}(\nu)}^p \right] < \infty.
$$

Moreover, for $\gamma \in (0,1/4)$, $\mathcal{Z}^{(0,3,M,N)}$ is γ-Hölder continuous in time on $[0,T]$ for any $M, N \in \mathbb{N}$ almost surely, and

$$
\sup_{M,N \in \mathbb{N}} E\left[\left(\sup_{s,t \in [0,T]; s \neq t} \frac{\left\| \mathcal{Z}_t^{(0,3,M,N)} - \mathcal{Z}_s^{(0,3,M,N)} \right\|_{L^p(\nu)}}{(t-s)^\gamma} \right)^p \right] < \infty.
$$

(iii) For $k = 2, 3$ and each M, $\{2^{-\alpha N} \mathcal{Z}^{(2,k,M,N)}; N \in \mathbb{N}\}$ converges almost surely in $C([0,\infty); B_{p,p}^{\alpha+1-(k/2)-\varepsilon}(\nu))$, and it holds that

$$\sup_{M,N\in\mathbb{N}} 2^{-p\alpha N} E\left[\sup_{t\in[0,T]} \left\| \mathcal{Z}_t^{(2,k,M,N)} \right\|^p_{B_{p,p}^{\alpha+1-(k/2)-\varepsilon}(\nu)}\right] < \infty.$$

(iv) For $k \in \mathbb{N}$, it holds that

$$\sup_{M,N\in\mathbb{N}} 2^{-p\alpha N} E\left[\sup_{t\in[0,T]} \left\| \mathcal{Z}_t^{(1,M,N)} \left(P_{M,N} \mathcal{Z}_t^{(0,3,M,N)}\right)^k \right\|^p_{B_{p,p}^{\alpha-(1/2)-\varepsilon}(\nu)}\right] < \infty.$$

(v) It holds that

$$2^{-p\alpha N} E\left[\sup_{t\in[0,T]} \left\| \widehat{\mathcal{Z}}_t^{(2,3,M,N)} \right\|^p_{B_{p,p}^{\alpha-1/2-\varepsilon}(\nu)}\right] \le C\left(1 + M^\sigma 2^{-N}\right)^p,$$

where C is a constant independent of M and N.

PROOF. First, we prove (i). The definition of Besov spaces implies

$$\begin{aligned} &E\left[\left\| \mathcal{Z}_t^{(k,N)} - \mathcal{Z}_s^{(k,N)} \right\|^2_{B_{2,2}^{\alpha-k/2-\varepsilon}(\nu)}\right] \\ &= \sum_{j=-1}^{\infty} 2^{2\alpha-(k+2\varepsilon)j} E\left[\left\| \Delta_j \left(\mathcal{Z}_t^{(k,N)} - \mathcal{Z}_s^{(k,N)}\right) \right\|^2_{L^2(\nu)}\right]. \end{aligned}$$

From this inequality and the translation invariance of the law of $\mathcal{Z}^{(k,N)}$ and the assumption that $\nu \in L^1(dx)$, we have

$$\begin{aligned} &2^{-2\alpha N} E\left[\left\| \mathcal{Z}_t^{(k,N)} - \mathcal{Z}_s^{(k,N)} \right\|^2_{B_{2,2}^{\alpha-k/2-\varepsilon}(\nu)}\right] \\ &\le C \sum_{j=-1}^{\infty} 2^{2\alpha(j-N)} 2^{-(k+2\varepsilon)j} E\left[\left\langle \Delta_j \left(\mathcal{Z}_t^{(k,N)} - \mathcal{Z}_s^{(k,N)}\right), \delta_0 \right\rangle^2\right] \\ &= C \sum_{j=-1}^{\infty} 2^{2\alpha(j-N)} 2^{-(k+2\varepsilon)j} E\left[\left\langle \mathcal{Z}_t^{(k,N)} - \mathcal{Z}_s^{(k,N)}, \Delta_j \delta_0 \right\rangle^2\right]. \end{aligned}$$

Since $\mathcal{Z}_t^{(k,N)}$ is given by the Wick polynomials of $P_N Z_t$, Lemma 3.1 and Plancherel theorem imply

$$\begin{aligned} E\left[\mathcal{Z}_t^{(k,N)}(x) \mathcal{Z}_s^{(k,N)}(y)\right] &= k! E\left[P_N Z_t(x) P_N Z_s(y)\right]^k \\ &= k! \langle V_N(t-s)\delta_x, \delta_y \rangle^k \\ &= k! \langle [2(m_0^2 - \triangle)]^{-1} e^{(t-s)(\triangle - m_0^2)} (P_N \delta_x), (P_N \delta_y) \rangle^k. \end{aligned}$$

Hence, from the fact that $\Delta_j \delta_0 = (2\pi)^{-3/2}\mathcal{F}^{-1}\varphi_j$, it holds that

$$
\begin{aligned}
&2^{-2\alpha N} E\left[\left\|\mathcal{Z}_t^{(k,N)} - \mathcal{Z}_s^{(k,N)}\right\|^2_{B_{2,2}^{\alpha-k/2-\varepsilon}(\nu)}\right] \\
&\leq C \sum_{j=-1}^{\infty} 2^{2\alpha(j-N)} 2^{-(k+2\varepsilon)j} \int_{\mathbb{R}^3}\int_{\mathbb{R}^3} \left[\left\langle [2(m_0^2 - \triangle)]^{-1}(P_N\delta_x), (P_N\delta_y)\right\rangle^k\right. \\
&\qquad \left. - \left\langle [2(m_0^2 - \triangle)]^{-1} e^{(t-s)(\triangle - m_0^2)}(P_N\delta_x), (P_N\delta_y)\right\rangle^k\right] \Delta_j\delta_0(x)\Delta_j\delta_0(y)dxdy \\
&= C \sum_{j=-1}^{\infty} 2^{2\alpha(j-N)} 2^{-(k+2\varepsilon)j} \\
&\quad \times \int_{\mathbb{R}^3} \cdots \int_{\mathbb{R}^3} \left[\left(I - e^{(t-s)(\triangle_{z_1} - m_0^2)} \cdots e^{(t-s)(\triangle_{z_k} - m_0^2)}\right)\right. \\
&\quad [2(m_0^2 - \triangle_{z_1})]^{-1} \cdots [2(m_0^2 - \triangle_{z_k})]^{-1} \left.\int_{\mathbb{R}^3} (\Delta_j\delta_0)(x)(P_N\delta_x)(z_1)\cdots(P_N\delta_x)(z_k)dx\right] \\
&\quad \times \left[\int_{\mathbb{R}^3} (\Delta_j\delta_0)(y)(P_N\delta_y)(z_1)\cdots(P_N\delta_y)(z_k)dy\right] dz_1\cdots dz_k
\end{aligned}
$$

where $\triangle_{z_l}$ is the Laplacian acting on functions with respect to the parameter z_l. Since

$$
\begin{aligned}
&I - e^{(t-s)(\triangle_{z_1} - m_0^2)} \cdots e^{(t-s)(\triangle_{z_k} - m_0^2)} \\
&= \sum_{l=1}^{k} \left(I - e^{(t-s)(\triangle_{z_l} - m_0^2)}\right) e^{(t-s)(\triangle_{z_{l+1}} - m_0^2)} \cdots e^{(t-s)(\triangle_{z_k} - m_0^2)},
\end{aligned}
$$

by Proposition 2.4 we have
(3.4)

$$
\begin{aligned}
&2^{-2\alpha N} E\left[\left\|\mathcal{Z}_t^{(k,N)} - \mathcal{Z}_s^{(k,N)}\right\|^2_{B_{2,2}^{\alpha-k/2-\varepsilon}(\nu)}\right] \\
&\quad \leq C(t-s)^{\varepsilon/2} \sum_{j=-1}^{\infty} 2^{2\alpha(j-N)} 2^{-(k+2\varepsilon)j} \\
&\qquad \times \left\| \int_{\mathbb{R}^3} (\Delta_j\delta_0)(x) \overbrace{(P_N\delta_x) \otimes \cdots \otimes (P_N\delta_x)}^{k} dx \right\|^2_{\underbrace{B_{2,2}^{-1+(\varepsilon/2)} \times B_{2,2}^{-1} \times \cdots \times B_{2,2}^{-1}}_{k}}.
\end{aligned}
$$

An explicit calculation implies

$$
(\Delta_{j_1} P_N\delta_x)(z_1) = \frac{1}{(2\pi)^3} \int_{\mathbb{R}^3} \varphi_{j_1}(\xi_1)\psi_N(\xi_1) e^{-\sqrt{-1}x\cdot\xi_1} e^{\sqrt{-1}z_1\cdot\xi} d\xi_1.
$$

Hence, it holds that

$$\begin{aligned}
&\int_{\mathbb{R}^3} (\Delta_j \delta_0)(x)(\Delta_{j_1} P_N \delta_x)(z_1) \cdots (\Delta_{j_k} P_N \delta_x)(z_k) dx \\
&= \frac{1}{(2\pi)^{3k}} \int_{\mathbb{R}^3} \cdots \int_{\mathbb{R}^3} \left(\prod_{l=1}^k \varphi_{j_l}(\xi_l) \psi_N(\xi_l) e^{\sqrt{-1} z_l \cdot \xi_l} \right) \\
&\qquad \times \left(\int_{\mathbb{R}^3} (\Delta_j \delta_0)(x) e^{-\sqrt{-1} x \cdot (\xi_1 + \cdots + \xi_k)} dx \right) d\xi_1 \cdots d\xi_k \\
&= \frac{1}{(2\pi)^{3k+(3/2)}} \int_{\mathbb{R}^3} \cdots \int_{\mathbb{R}^3} \left(\prod_{l=1}^k \varphi_{j_l}(\xi_l) \psi_N(\xi_l) e^{\sqrt{-1} z_l \cdot \xi_l} \right) \\
&\qquad \times \left(\int_{\mathbb{R}^3} (\mathcal{F}^{-1} \varphi_j)(x) e^{-\sqrt{-1} x \cdot (\xi_1 + \cdots + \xi_k)} dx \right) d\xi_1 \cdots d\xi_k \\
&= \frac{1}{(2\pi)^{3k}} \int_{\mathbb{R}^3} \cdots \int_{\mathbb{R}^3} \left(\prod_{l=1}^k \varphi_{j_l}(\xi_l) \psi_N(\xi_l) \right) \\
&\qquad \times \varphi_j(\xi_1 + \cdots + \xi_k) e^{\sqrt{-1}(z_1 \cdot \xi_1 + \cdots + z_k \cdot \xi_k)} d\xi_1 \cdots d\xi_k.
\end{aligned}$$

This formula implies that

$$\begin{aligned}
&\left\| \int_{\mathbb{R}^3} (\Delta_j \delta_0)(x) \overbrace{(P_N \delta_x) \otimes \cdots \otimes (P_N \delta_x)}^{k} dx \right\|^2_{\overbrace{B_{2,2}^{-1+(\varepsilon/2)} \times B_{2,2}^{-1} \times \cdots \times B_{2,2}^{-1}}^{k}} \\
&= \frac{1}{(2\pi)^{3k}} \sum_{j_1, \cdots, j_k = -1}^{\infty} 2^{-(2-\varepsilon) j_1} 2^{-2 j_2} \cdots 2^{-2 j_k} \\
&\qquad \times \int_{\mathbb{R}^3} \cdots \int_{\mathbb{R}^3} \left(\prod_{l=1}^k \varphi_{j_l}(\xi_l)^2 \psi_N(\xi_l)^2 \right) \varphi_j(\xi_1 + \cdots + \xi_k)^2 d\xi_1 \cdots d\xi_k.
\end{aligned}$$

By this equality, (3.4) and the fact that $\varphi_{j_l}(\xi_l)\psi_N(\xi_l) = 0$ for $j_l > N+3$, we have

$$\begin{aligned}
&2^{-2\alpha N} E\left[\left\| \mathcal{Z}_t^{(k,N)} - \mathcal{Z}_s^{(k,N)} \right\|^2_{B_{2,2}^{\alpha - k/2 - \varepsilon}(\nu)} \right] \\
&\le C(t-s)^{\varepsilon/2} \sum_{j=-1}^{\infty} 2^{2\alpha(j-N)} 2^{-(k+2\varepsilon)j} \sum_{j_1, \cdots, j_k = -1}^{N+3} 2^{-(2-\varepsilon) j_1} 2^{-2 j_2} \cdots 2^{-2 j_k} \\
&\qquad \times \int_{\mathbb{R}^3} \cdots \int_{\mathbb{R}^3} \varphi_{j_1}(\xi_1)^2 \cdots \varphi_{j_k}(\xi_k)^2 \varphi_j(\xi_1 + \cdots + \xi_k)^2 d\xi_1 \cdots d\xi_k.
\end{aligned}$$

When $j \ge N+6+k$, $j_1, \dots j_m \le N+3$ and $|2^{-j_l}\xi_l| \le 8/3$ for $l = 1, \dots, k$, then

$$|2^{-j}\xi_1 + \cdots + 2^{-j}\xi_k| \le 2^{-3-k} \cdot k \cdot \frac{8}{3} < \frac{3}{4}.$$

Hence, again in view of the support of φ, it holds that

$$\begin{aligned}
&2^{-2\alpha N} E\left[\left\|\mathcal{Z}_t^{(k,N)}-\mathcal{Z}_s^{(k,N)}\right\|^2_{B^{\alpha-k/2-\varepsilon}_{2,2}(\nu)}\right]\\
&\le C(t-s)^{\varepsilon/2}\sum_{j=-1}^{N+5+k}2^{-(k+2\varepsilon)j}\sum_{j_1,\cdots,j_k=-1}^{N+3}2^{-(2-\varepsilon)j_1}2^{-2j_2}\cdots 2^{-2j_k}\\
&\qquad\times\int_{\mathbb{R}^3}\cdots\int_{\mathbb{R}^3}\varphi_{j_1}(\xi_1)^2\cdots\varphi_{j_k}(\xi_k)^2\varphi_j(\xi_1+\cdots+\xi_k)^2 d\xi_1\cdots d\xi_k.
\end{aligned}$$

Hence, we have

$$\begin{aligned}
&2^{-2\alpha N} E\left[\left\|\mathcal{Z}_t^{(k,N)}-\mathcal{Z}_s^{(k,N)}\right\|^2_{B^{\alpha-k/2-\varepsilon}_{2,2}(\nu)}\right]\\
&\le C(t-s)^{\varepsilon/2}\sum_{j_1,\cdots,j_k=-1}^{N+3}2^{-(2-\varepsilon)j_1}2^{-2j_2}\cdots 2^{-2j_k}\\
&\quad\times\int_{\mathbb{R}^3}\cdots\int_{\mathbb{R}^3}\left(2^{j_1}|\xi_1|^{-1}\right)^{2-\varepsilon}\left(2^{j_2}|\xi_2|^{-1}\right)^2\cdots\left(2^{j_k}|\xi_k|^{-1}\right)^2\\
&\quad\times\left(1+|\xi_1+\cdots+\xi_k|^2\right)^{-[(k/2)+\varepsilon]}\varphi_{j_1}(\xi_1)^2\cdots\varphi_{j_k}(\xi_k)^2\\
&\qquad\qquad\qquad\times\chi(\xi_1+\cdots+\xi_k)^2 d\xi_1\cdots d\xi_k\\
&\quad+C(t-s)^{\varepsilon/2}\sum_{j=0}^{N+5+k}2^{-(k+2\varepsilon)j}\sum_{j_1,\cdots,j_k=-1}^{N+3}2^{-(2-\varepsilon)j_1}2^{-2j_2}\cdots 2^{-2j_k}\\
&\quad\times\int_{\mathbb{R}^3}\cdots\int_{\mathbb{R}^3}\left(2^{j_1}|\xi_1|^{-1}\right)^{2-\varepsilon}\left(2^{j_2}|\xi_2|^{-1}\right)^2\cdots\left(2^{j_k}|\xi_k|^{-1}\right)^2\\
&\quad\times\left(2^{j}|\xi_1+\cdots+\xi_k|^{-1}\right)^{k+2\varepsilon}\varphi_{j_1}(\xi_1)^2\cdots\varphi_{j_k}(\xi_k)^2\varphi_j(\xi_1+\cdots+\xi_k)^2 d\xi_1\cdots d\xi_k\\
&\le C(t-s)^{\varepsilon/2}\int_{\mathbb{R}^3}\cdots\int_{\mathbb{R}^3}\frac{|\xi_1|^{\varepsilon}}{|\xi_1|^2\cdots|\xi_k|^2(1+|\xi_1+\cdots+\xi_k|^2)^{(k/2)+\varepsilon}}d\xi_1\cdots d\xi_k.
\end{aligned}$$

Thus, for $k=1,2$ we obtain

$$2^{-2\alpha N} E\left[\left\|\mathcal{Z}_t^{(k,N)}-\mathcal{Z}_s^{(k,N)}\right\|^2_{B^{\alpha-k/2-\varepsilon}_{2,2}(\nu)}\right]\le C(t-s)^{\varepsilon/2}.$$

This inequality and the hypercontractivity of Gaussian polynomials (see e.g. [**NP12**, Theorem 2.7.2]) yield

$$2^{-p\alpha N} E\left[\left\|\mathcal{Z}_t^{(k,N)}-\mathcal{Z}_s^{(k,N)}\right\|^p_{B^{\alpha-k/2-\varepsilon}_{p,p}(\nu)}\right]\le C(t-s)^{\varepsilon p/2}. \tag{3.5}$$

In view of Lemma 3.1(iii), by a similar calculation we have

$$\sup_{t\in[0,T]}\sum_{N=1}^{\infty}2^{-\alpha N}E\left[\left\|\mathcal{Z}_t^{(k,N+1)}-\mathcal{Z}_t^{(k,N)}\right\|^p_{B^{\alpha-k/2-\varepsilon}_{p,p}(\nu)}\right]^{1/p}<\infty. \tag{3.6}$$

Indeed, similarly to the proof of Lemma 3.1 we can prove that

$$E\left[\langle f,P_{N+1}Z_t\rangle\langle g,P_N Z_t\rangle\right]=\left\langle P_N P_{N+1}[2(m_0^2-\triangle)]^{-1}f,g\right\rangle,$$

and hence

$$
\begin{aligned}
&E\left[\left\langle \mathcal{Z}_t^{(k,N+1)} - \mathcal{Z}_t^{(k,N)}, \Delta_j\delta_0\right\rangle^2\right] \\
&= k!\int_{\mathbb{R}^3}\int_{\mathbb{R}^3}\Big[\left\langle [2(m_0^2-\triangle)]^{-1}P_{N+1}\delta_x, P_{N+1}\delta_y\right\rangle^k + \left\langle [2(m_0^2-\triangle)]^{-1}P_N\delta_x, P_N\delta_y\right\rangle^k \\
&\qquad\qquad -2\left\langle [2(m_0^2-\triangle)]^{-1}P_{N+1}\delta_x, P_N\delta_y\right\rangle^k\Big]\Delta_j\delta_0(x)\Delta_j\delta_0(y)dxdy \\
&= k!\int_{\mathbb{R}^3}\cdots\int_{\mathbb{R}^3}\Big[[2(m_0^2-\triangle_{z_1})]^{-1}\cdots[2(m_0^2-\triangle_{z_k})]^{-1} \\
&\qquad \sum_{m=1}^k\int_{\mathbb{R}^3}(\Delta_j\delta_0)(x)(P_N\delta_x)(z_1)\cdots(P_N\delta_x)(z_{m-1}) \\
&\qquad\quad \times[(P_{N+1}-P_N)\delta_x](z_m)(P_{N+1}\delta_x)(z_{m+1})\cdots(P_{N+1}\delta_x)(z_k)dx\Big] \\
&\qquad \times\sum_{\tilde{m}=1}^k\int_{\mathbb{R}^3}(\Delta_j\delta_0)(y)(P_N\delta_y)(z_1)\cdots(P_N\delta_y)(z_{\tilde{m}-1}) \\
&\qquad\quad \times[(P_{N+1}-P_N)\delta_y](z_{\tilde{m}})(P_{N+1}\delta_y)(z_{\tilde{m}+1})\cdots(P_{N+1}\delta_y)(z_k)dy\Big]dz_1\cdots dz_k \\
&\le C\sum_{m=1}^k\Bigg\|\int_{\mathbb{R}^3}(\Delta_j\delta_0)(x)\overbrace{(P_N\delta_x)\otimes\cdots\otimes(P_N\delta_x)}^{m-1}\otimes[(P_{N+1}-P_N)\delta_x] \\
&\qquad\qquad \otimes\overbrace{(P_{N+1}\delta_x)\otimes\cdots\otimes(P_{N+1}\delta_x)}^{k-m}dx\Bigg\|^2_{\overbrace{B^{-1}_{2,2}\times B^{-1}_{2,2}\times\cdots\times B^{-1}_{2,2}}^{k}}.
\end{aligned}
$$

Similarly to above, from this inequality and Lemma 3.1 we have (3.6). Applying the argument in the proof of [**AK20**, Proposition 3.3], from (3.5) and (3.6) we obtain the almost-sure convergence of $\{2^{-\alpha N}\mathcal{Z}^{(k,N)}; N\in\mathbb{N}\}$ in $C([0,\infty); B_{p,p}^{\alpha-k/2-\varepsilon}(\nu))$. Thus, we get (i).

We have (ii) in a similar way, because of the facts that

$$
\begin{aligned}
&E\left[\left(\Delta_i\mathcal{Z}_t^{(0,k,M,N)}\right)(x)\left(\Delta_j\mathcal{Z}_s^{(0,k,M,N)}\right)(y)\right] \\
&= \int_{-\infty}^t\int_{-\infty}^s E\Big[\left\langle \mathcal{Z}_u^{(k,M,N)}, P_{M,N}e^{(t-u)(\triangle-m_0^2)}\Delta_i\delta_x\right\rangle \\
&\qquad\qquad \times\left\langle \mathcal{Z}_v^{(k,M,N)}, P_{M,N}e^{(s-v)(\triangle-m_0^2)}\Delta_j\delta_y\right\rangle\Big]dudv \\
&= k!\int_{-\infty}^t\int_{-\infty}^s\int_{\mathbb{R}^3}\int_{\mathbb{R}^3}\langle V_N(|u-v|)\delta_{z_1},\delta_{z_2}\rangle^k\left(\rho_M^kP_{M,N}e^{(t-u)(\triangle-m_0^2)}\Delta_i\delta_x\right)(z_1) \\
&\qquad\qquad \times\left(\rho_M^kP_{M,N}e^{(s-v)(\triangle-m_0^2)}\Delta_j\delta_y\right)(z_2)dz_1dz_2dudv \\
&= k!\int_0^\infty\int_0^\infty\int_{\mathbb{R}^3}\int_{\mathbb{R}^3}\langle V_N(|(t-s)-(\tilde{u}-\tilde{v})|)\delta_{z_1},\delta_{z_2}\rangle^k \\
&\quad \times\left(\rho_M^{k+1}e^{\tilde{u}(\triangle-m_0^2)}(\Delta_iP_N\delta_x)\right)(z_1)\left(\rho_M^{k+1}e^{\tilde{v}(\triangle-m_0^2)}(\Delta_jP_N\delta_y)\right)(z_2)dz_1dz_2d\tilde{u}d\tilde{v}.
\end{aligned}
$$

and that

$$\begin{aligned}&\int_{\mathbb{R}^3}\left(\rho_M^{k+1}e^{\tilde{u}(\triangle-m_0^2)}(\Delta_i P_N\delta_x)\right)(z_1)e^{-\sqrt{-1}z_1\cdot\xi}dz_1\\&=\frac{1}{(2\pi)^{3/2}}\int_{\mathbb{R}^3}(\mathcal{F}\rho_M^{k+1})(\xi-\eta)\left(e^{-\tilde{u}(\eta^2+m_0^2)}\varphi_i(\eta)\psi_N(\eta)e^{-\sqrt{-1}\eta\cdot x}\right)d\eta.\end{aligned}$$

For the estimate of the Hölder continuity of $\mathcal{Z}_t^{(0,3,M,N)}$, we show

$$E\left[\left\|\mathcal{Z}_t^{(0,3,M,N)}-\mathcal{Z}_s^{(0,3,M,N)}\right\|_{L^p(\nu)}^p\right]\le C(t-s)^{p(\gamma+\varepsilon)}$$

for sufficiently small $\varepsilon>0$ and sufficiently large p in a similar way as above, and apply the Kolmogorov continuity theorem.

Next we prove (iii) for $k=2$. However, it is very complicated and difficult to write down all details explicitly. So, we give only a sketch of the proof. In view of the above argument of hypercontractivity, it is sufficient to show that

$$2^{-2\alpha N}E\left[\left\|\mathcal{Z}_t^{(2,2,M,N)}-\mathcal{Z}_s^{(2,2,M,N)}\right\|_{B_{2,2}^{\alpha-\varepsilon}(\nu)}^2\right]\le C(t-s)^{\varepsilon/2}\tag{3.7}$$

$$\sum_{N=1}^{\infty}2^{-2\alpha N}E\left[\left\|\mathcal{Z}_t^{(2,2,M,N+1)}-\mathcal{Z}_t^{(2,2,M,N)}\right\|_{B_{2,2}^{\alpha-\varepsilon}(\nu)}^2\right]\le C,\tag{3.8}$$

where C is a constant independent of M and N. By the definition of $\mathcal{Z}^{(2,2,M,N)}$ and Proposition 2.5

$$\begin{aligned}&E\left[\left\|\mathcal{Z}_t^{(2,2,M,N)}-\mathcal{Z}_s^{(2,2,M,N)}\right\|_{B_{2,2}^{\alpha-\varepsilon}(\nu)}^2\right]\\&\le CE\left[\left\|\left(\mathcal{Z}_t^{(2,M,N)}-\mathcal{Z}_s^{(2,M,N)}\right)\circleddash\int_{-\infty}^{s}e^{(s-u)(\triangle-m_0^2)}P_N^2\mathcal{Z}_u^{(2,M,N)}du\right\|_{B_{2,2}^{\alpha-\varepsilon}(\nu)}^2\right]\\&\quad+CE\left[\left\|\mathcal{Z}_t^{(2,M,N)}\right.\right.\\&\qquad\left.\left.\circleddash\int_{-\infty}^{s}\left[e^{(t-s)(\triangle-m_0^2)}-I\right]e^{(s-u)(\triangle-m_0^2)}P_N^2\mathcal{Z}_u^{(2,M,N)}du\right\|_{B_{2,2}^{\alpha-\varepsilon}(\nu)}^2\right]\\&\quad+CE\left[\left\|\mathcal{Z}_t^{(2,M,N)}\circleddash\int_{s}^{t}e^{(t-u)(\triangle-m_0^2)}P_N^2\mathcal{Z}_u^{(2,M,N)}du\right\|_{B_{2,2}^{\alpha-\varepsilon}(\nu)}^2\right].\end{aligned}$$

We have now to estimate all terms on the right-hand side. However, we consider only the last term, because the others are similarly obtained. By the definitions of Besov spaces and paraproducts, commutativity of the heat semigroup and Δ_j, and the supports of φ and ψ, we have

$$\begin{aligned}
&2^{-2\alpha N} \left\| \mathcal{Z}_t^{(2,M,N)} \circleddash \int_s^t e^{(t-u)(\triangle - m_0^2)} P_N^2 \mathcal{Z}_u^{(2,M,N)} du \right\|_{B_{2,2}^{\alpha-\varepsilon}(\nu)}^2 \\
&= 2^{-2\alpha N} \sum_{l=-1}^{\infty} 2^{2(\alpha-\varepsilon)l} \\
&\quad \times \left\| \Delta_l \left[\sum_{i,j;|i-j|\leq 1} (\Delta_i \mathcal{Z}_t^{(2,M,N)}) \left(\Delta_j \int_s^t e^{(t-u)(\triangle - m_0^2)} P_N^2 \mathcal{Z}_u^{(2,M,N)} du \right) \right] \right\|_{L^2(\nu)}^2 \\
&\leq C \sum_{l=-1}^{\infty} 2^{-2\varepsilon l} \sum_{i_1,j_1;|i_1-j_1|\leq 1} \sum_{i_2,j_2;|i_2-j_2|\leq 1} \\
&\quad \times \int_{\mathbb{R}^3} \int_{\mathbb{R}^3} \Delta_l \left[(\Delta_{i_1} \mathcal{Z}_t^{(2,M,N)}) \left(\int_s^t e^{(t-u)(\triangle - m_0^2)} P_N^2 \Delta_{j_1} \mathcal{Z}_u^{(2,M,N)} du \right) \right] (x) \\
&\quad \times \Delta_l \left[(\Delta_{i_2} \mathcal{Z}_t^{(2,M,N)}) \left(\int_s^t e^{(t-v)(\triangle - m_0^2)} P_N^2 \Delta_{j_2} \mathcal{Z}_v^{(2,M,N)} dv \right) \right] (y) \nu(x) \nu(y) dx dy
\end{aligned}$$

for some constant C. Hence, we need to estimate

$$\begin{aligned}
E \Big[&\left\langle \mathcal{Z}_t^{(2,M,N)}, \Delta_{i_1} \delta_x \right\rangle \left\langle \mathcal{Z}_u^{(2,M,N)}, e^{(t-u)(\triangle - m_0^2)} P_N^2 \Delta_{j_1} \delta_x \right\rangle \\
&\times \left\langle \mathcal{Z}_t^{(2,M,N)}, \Delta_{i_2} \delta_y \right\rangle \left\langle \mathcal{Z}_v^{(2,M,N)}, e^{(t-v)(\triangle - m_0^2)} P_N^2 \Delta_{j_2} \delta_y \right\rangle \Big].
\end{aligned}$$

This term is estimated by the product formula (see Theorem 2.7.10 in [**NP12**]) or by the calculation of the expectation of Gaussian polynomials by pairing (see Theorem 1.28 in [**Jan97**]). The rest is similar to that of (i), and by combining the estimates above we obtain (3.7). We also have (3.8) by repeating a similar argument. We obtain the case that $k = 3$ in a similar way, however the calculation is more complicated. So, we skip giving details.

Next we show ((iv)). We fix $\varepsilon \in (0, 1/8]$ and $\alpha \in [0, \infty)$. First we show

$$\sup_{M,N\in\mathbb{N}} 2^{-p\alpha N} E \left[\sup_{t\in[0,T]} \left\| \mathcal{Z}_t^{(1,M,N)} \circleddash \left(P_{M,N} \mathcal{Z}_t^{(0,3,M,N)} \right)^k \right\|_{B_{p,p}^{\alpha-\varepsilon}(\nu)}^p \right] < \infty \tag{3.9}$$

for $p \in [1, \infty)$ by induction in k. Similarly to the proof of (iii) we have (3.9) in the case that $k = 1$. Assume that $k \geq 2$ and that (3.9) holds for $k - 1$. We decompose the product as

$$
\begin{aligned}
&\mathcal{Z}_t^{(1,M,N)} \circleddash \left(P_{M,N}\mathcal{Z}_t^{(0,3,M,N)}\right)^k \\
&= \mathcal{Z}_t^{(1,M,N)} \circleddash \left[\left(P_{M,N}\mathcal{Z}_t^{(0,3,M,N)}\right) \circleddash \left(P_{M,N}\mathcal{Z}_t^{(0,3,M,N)}\right)^{k-1}\right] \\
&\quad + \mathcal{Z}_t^{(1,M,N)} \circleddash \left[\left(P_{M,N}\mathcal{Z}_t^{(0,3,M,N)}\right) \circledless \left(P_{M,N}\mathcal{Z}_t^{(0,3,M,N)}\right)^{k-1}\right] \\
&\qquad\qquad - \left(P_{M,N}\mathcal{Z}_t^{(0,3,M,N)}\right)\left[\mathcal{Z}_t^{(1,M,N)} \circleddash \left(P_{M,N}\mathcal{Z}_t^{(0,3,M,N)}\right)^{k-1}\right] \\
&\quad + \left(P_{M,N}\mathcal{Z}_t^{(0,3,M,N)}\right)\left[\mathcal{Z}_t^{(1,M,N)} \circleddash \left(P_{M,N}\mathcal{Z}_t^{(0,3,M,N)}\right)^{k-1}\right] \\
&\quad + \mathcal{Z}_t^{(1,M,N)} \circleddash \left[\left(P_{M,N}\mathcal{Z}_t^{(0,3,M,N)}\right) \circledgtr \left(P_{M,N}\mathcal{Z}_t^{(0,3,M,N)}\right)^{k-1}\right] \\
&\qquad\qquad - \left(P_{M,N}\mathcal{Z}_t^{(0,3,M,N)}\right)^{k-1}\left[\mathcal{Z}_t^{(1,M,N)} \circleddash \left(P_{M,N}\mathcal{Z}_t^{(0,3,M,N)}\right)\right] \\
&\quad + \left(P_{M,N}\mathcal{Z}_t^{(0,3,M,N)}\right)^{k-1}\left[\mathcal{Z}_t^{(1,M,N)} \circleddash \left(P_{M,N}\mathcal{Z}_t^{(0,3,M,N)}\right)\right].
\end{aligned}
$$

Then, by Propositions 2.5 and 2.13 we have, for some constant $C > 0$

$$
\begin{aligned}
&\left\|\mathcal{Z}_t^{(1,M,N)} \circleddash \left(P_{M,N}\mathcal{Z}_t^{(0,3,M,N)}\right)^k\right\|_{B_{p,p}^{\alpha-\varepsilon}(\nu)} \\
&\le C\left\|\mathcal{Z}_t^{(1,M,N)}\right\|_{B_{2p,2p}^{\alpha-(1+\varepsilon)/2}(\nu)} \\
&\qquad\qquad \times \left\|\left(P_{M,N}\mathcal{Z}_t^{(0,3,M,N)}\right) \circleddash \left(P_{M,N}\mathcal{Z}_t^{(0,3,M,N)}\right)^{k-1}\right\|_{B_{2p,2p}^{1/2+\varepsilon}(\nu)} \\
&\quad + C\left\|\mathcal{Z}_t^{(1,M,N)}\right\|_{B_{2p,2p}^{\alpha-1/2-\varepsilon}(\nu)} \left\|\left(P_{M,N}\mathcal{Z}_t^{(0,3,M,N)}\right)\right\|_{B_{4p,4p}^{1/2-\varepsilon}(\nu)} \\
&\qquad\qquad\qquad\qquad \times \left\|\left(P_{M,N}\mathcal{Z}_t^{(0,3,M,N)}\right)^{k-1}\right\|_{B_{4p,4p}^{2\varepsilon}(\nu)} \\
&\quad + C\left\|P_{M,N}\mathcal{Z}_t^{(0,3,M,N)}\right\|_{L^{2p}(\nu)} \left\|\mathcal{Z}_t^{(1,M,N)} \circleddash \left(P_{M,N}\mathcal{Z}_t^{(0,3,M,N)}\right)^{k-1}\right\|_{B_{2p,2p}^{\alpha-\varepsilon}(\nu)} \\
&\quad + C\left\|\left(P_{M,N}\mathcal{Z}_t^{(0,3,M,N)}\right)^{k-1}\right\|_{L^{2p}(\nu)} \left\|\mathcal{Z}_t^{(1,M,N)} \circleddash \left(P_{M,N}\mathcal{Z}_t^{(0,3,M,N)}\right)\right\|_{B_{2p,2p}^{\alpha-\varepsilon}(\nu)}.
\end{aligned}
$$

In view of the induction assumption, Proposition 2.5, (i) and (ii), we see that (3.9) also holds for k. Since

$$
\begin{aligned}
&\mathcal{Z}_t^{(1,M,N)} \left(P_{M,N}\mathcal{Z}_t^{(0,3,M,N)}\right)^k \\
&= \mathcal{Z}_t^{(1,M,N)} \circledless \left(P_{M,N}\mathcal{Z}_t^{(0,3,M,N)}\right)^k + \mathcal{Z}_t^{(1,M,N)} \circledgtr \left(P_{M,N}\mathcal{Z}_t^{(0,3,M,N)}\right)^k \\
&\quad + \mathcal{Z}_t^{(1,M,N)} \circleddash \left(P_{M,N}\mathcal{Z}_t^{(0,3,M,N)}\right)^k,
\end{aligned}
$$

by applying Proposition 2.5, (i), (ii) and (3.9), we have (iv).

Finally we show (v). Decompose $\widehat{\mathcal{Z}}_t^{(2,3,M,N)}$ as

$$
\begin{aligned}
\widehat{\mathcal{Z}}_t^{(2,3,M,N)} &= \rho_M^2 \mathcal{Z}_t^{(2,3,M,N)} \\
&\quad + \left(P_{M,N} \int_{-\infty}^t e^{(t-s)(\triangle - m_0^2)} P_N \left(\rho_M \circledgreater \mathcal{Z}_s^{(3,M,N)} \right) ds \right) \circleddash \mathcal{Z}_t^{(2,M,N)} \\
&\quad + \left(\rho_M \left[\int_{-\infty}^t e^{(t-s)(\triangle - m_0^2)} P_N^2 \left(\rho_M \circledless \mathcal{Z}_s^{(3,M,N)} \right) ds \right.\right. \\
&\qquad\qquad \left.\left. - \left(\rho_M \circledless \int_{-\infty}^t e^{(t-s)(\triangle - m_0^2)} P_N^2 \mathcal{Z}_s^{(3,M,N)} ds \right) \right] \right) \circleddash \mathcal{Z}_t^{(2,M,N)} \\
&\quad + \left[\rho_M \left(\rho_M \circledless \int_{-\infty}^t e^{(t-s)(\triangle - m_0^2)} P_N^2 \mathcal{Z}_s^{(3,M,N)} ds \right) \right] \circleddash \mathcal{Z}_t^{(2,M,N)} \\
&\qquad - \rho_M^2 \left[\left(\int_{-\infty}^t e^{(t-s)(\triangle - m_0^2)} P_N^2 \mathcal{Z}_s^{(3,M,N)} ds \right) \circleddash \mathcal{Z}_t^{(2,M,N)} \right].
\end{aligned}
$$

Hence, we get the required estimate from this, (i), (ii), (iii), Propositions 2.5, 2.13 and 2.14, and the fact that $\sup_M \|\rho_M\|_{B^2_\infty(\nu)} < \infty$. □

REMARK 3.4. The case where $\alpha = 0$ in Proposition 3.3 is also discussed in [**HL18**] in the framework of the regularity structures.

CHAPTER 4

Approximation equations and their transformation

In this section, we will construct an invariant probability measure and a flow associated to the Φ_3^4-measure. We use the same notations as in Chapters 2 and 3.

Let $\lambda_0 \in (0, \infty)$ and $\lambda \in (0, \lambda_0]$ be fixed. Let $\nu(x) := (1 + |x|^2)^{-\sigma/2}$ and assume

$$9 < \sigma^2 < 2m_0^2. \tag{4.1}$$

Define a function $U_{M,N}$ and a measure $\mu_{M,N}$ as in Section 1.2. Recall that $\{\mu_{M,N}\}$ is an approximation sequence for the Φ_3^4-measure, which will be constructed below as an invariant probability measure of the flow associated with the stochastic quantization equation. For simplicity, let $C_*^{(M,N)}(x) := C_1^{(N)} - 3\lambda C_2^{(M,N)}(x)$.

Consider the stochastic partial differential equation on $\mathbb{R}^3$

$$\begin{aligned}
&\partial_t Y_t^{M,N}(x) \\
&\quad = \dot{W}_t(x) - (-\triangle + m_0^2) Y_t^{M,N}(x) \\
&\qquad - \lambda P_{M,N}^* \left\{ (P_{M,N} Y_t^{M,N})^3(x) - 3C_*^{(M,N)}(x) \rho_M(x)^2 P_{M,N} Y_t^{M,N}(x) \right\}
\end{aligned} \tag{4.2}$$

where $\dot{W}_t(x)$ is a white noise with parameter $(t, x) \in [0, \infty) \times \mathbb{R}^3$. First, we prove that this SPDE is associated to $\mu_{M,N}$, in the sense that $\mu_{M,N}$ is the invariant measure for $Y_t^{M,N}$.

THEOREM 4.1. *Let $\alpha \in (1/2, \infty)$ and $\varepsilon \in (0, \infty)$. For each M and N, (4.2) has a unique global mild solution as a continuous stochastic process on $B_2^{-\alpha-\varepsilon}(\nu)$ almost surely for any initial value $Y_0^{M,N} \in B_2^{-\alpha}(\nu)$ and the solution $Y_t^{M,N}$ satisfies*

$$E\left[\sup_{s \in [0,T]} \left\| Y_s^{M,N} \right\|_{B_2^{-\alpha-\varepsilon}(\nu)}^2 \right] < \infty.$$

Moreover, $\mu_{M,N}$ is the invariant measure for the approximation process $Y^{M,N}$.

PROOF. Consider the decomposition $Y_t^{M,N} = P_{2N} Y_t^{M,N} + P_{2N}^\perp Y_t^{M,N}$ where $P_{2N}^\perp := I - P_{2N}$. Then, $(P_{2N} Y_t^{M,N}, P_{2N}^\perp Y_t^{M,N})$ should solve the following system of coupled stochastic partial differential equations:

(4.3)

$$\begin{aligned}
\partial_t P_{2N} Y_t^{M,N} &= P_{2N} \dot{W}_t - (-\triangle + m_0^2) P_{2N} Y_t^{M,N} \\
&\quad - \lambda P_{M,N}^* \left\{ (P_{M,N} P_{2N} Y_t^{M,N})^3 - 3C_*^{(M,N)} \rho_M^2 P_{M,N} P_{2N} Y_t^{M,N} \right\}
\end{aligned}$$

(4.4)

$$\partial_t P_{2N}^\perp Y_t^{M,N} = P_{2N}^\perp \dot{W}_t - (-\triangle + m_0^2) P_{2N}^\perp Y_t^{M,N}.$$

It is easy to see that $\{P_{2N}\dot{W}_t; t \in [0,\infty)\}$ and $\{P_{2N}^{\perp}\dot{W}_t; t \in [0,\infty)\}$ are independent. Hence, it is sufficient to solve (4.3) and (4.4) independently.

The solution to (4.4) is obtained explicitly as

$$P_{2N}^{\perp}Y_t^{M,N} = e^{t(\triangle - m_0^2)}P_{2N}^{\perp}Y_0^{M,N} + \int_0^t e^{(t-s)(\triangle - m_0^2)}P_{2N}^{\perp}\dot{W}_s ds. \tag{4.5}$$

Hence, by explicit calculation, for $s,t \in [0,\infty)$ such that $s<t$

$$\begin{aligned}
&E\left[\left\|P_{2N}^{\perp}Y_t^{M,N} - P_{2N}^{\perp}Y_s^{M,N}\right\|_{B_{2,2}^{-\alpha-\varepsilon}(\nu)}^2\right] \\
&\le 2\left\|\left(e^{(t-s)(\triangle - m_0^2)} - I\right)P_{2N}^{\perp}Y_0^{M,N}\right\|_{B_{2,2}^{-\alpha-\varepsilon}(\nu)}^2 \\
&\qquad + 2E\left[\left\|\int_s^t e^{(t-u)(\triangle - m_0^2)}P_{2N}^{\perp}\dot{W}_u du\right\|_{B_{2,2}^{-\alpha-\varepsilon}(\nu)}^2\right] \\
&\le 2(t-s)^{2\varepsilon}\left\|Y_0^{M,N}\right\|_{B_{2,2}^{-\alpha}(\nu)}^2 \\
&\qquad + 2\sum_{j=-1}^{\infty} 2^{-2j(\alpha+\varepsilon)}E\left[\left\|\Delta_j\int_s^t e^{(t-u)(\triangle - m_0^2)}P_{2N}^{\perp}\dot{W}_u du\right\|_{L^2(\nu)}^2\right].
\end{aligned}$$

By another explicit calculation we have

$$\begin{aligned}
&E\left[\left\|\Delta_j\int_s^t e^{(t-u)(\triangle - m_0^2)}P_{2N}^{\perp}\dot{W}_u du\right\|_{L^2(\nu)}^2\right] \\
&\le \int_s^t\int_{\mathbb{R}^3}\int_{\mathbb{R}^3}\left([(\Delta_j P_{2N}^{\perp}p)(t-u,\cdot)](x-y)\right)^2\nu(x)dxdydu \\
&\le C\int_s^t\int_{\mathbb{R}^3}\varphi_j(|\xi|)^2 e^{-2(t-u)(|\xi|^2+m_0^2)}d\xi du \\
&\le C2^j(t-s)
\end{aligned}$$

where $p(t,x)$ is the heat kernel, i.e.

$$p(t,x) = (4\pi t)^{-3/2}e^{-|x|^2/4t}, \quad (t,x) \in (0,\infty)\times\mathbb{R}^3$$

and C is a constant depending on ν. These inequalities imply that $P_{2N}^{\perp}Y_t^{M,N}$ is a $B_2^{-\alpha-\varepsilon}(\nu)$-valued continuous process.

Now we construct a solution to (4.3) in $L^2(\nu)$. Since P_{2N} is a trace class operator in $L^2(dx)$ and $P_{M,N}$ is a bounded linear operator from $B_p^s(\nu)$ to $L^2(\nu)$ for each $p \in [1,\infty]$ and $s \in \mathbb{R}$, we have the pathwise uniqueness of the local solution $P_{2N}Y_t^{M,N}$ in time to (4.3) on $L^2(\nu)$ by well-known facts of the theory of stochastic partial differential equations with locally Lipschitz coefficients.

Next, we show the global existence of the solution to (4.3). Let

$$Z_t^Y := e^{t(\triangle - m_0^2)}Y_0^{M,N} + \int_0^t e^{(t-s)(\triangle - m_0^2)}\dot{W}_s ds, \quad Y_t^{M,N,(1)} := Y_t^{M,N} - Z_t^Y.$$

Let $T>0$. Define a stopping time $\tau_L := \min\{T, \inf\{t>0; \|P_{2N}Y_t^{M,N,(1)}\|_{L^2} > L\}\}$ for $L \in [0,\infty)$. Since

$$\partial_t Z_t^Y = \dot{W}_t + (\triangle - m_0^2)Z_t^Y, \quad Z_0^Y = Y_0^{M,N},$$

$Y_t^{M,N,(1)}$ satisfies that $Y_0^{M,N,(1)} = 0$ and

$$\begin{aligned}\partial_t P_{2N} Y_t^{M,N,(1)} &= -(-\triangle + m_0^2) P_{2N} Y_t^{M,N,(1)} \\ &\quad - \lambda P_{M,N}^* \left\{ (P_{M,N} P_{2N} Y_t^{M,N})^3 - 3C_*^{(M,N)} \rho_M^2 P_{M,N} P_{2N} Y_t^{M,N} \right\}.\end{aligned}$$

The mild form of the solution to this equation implies

$$\begin{aligned}P_{2N} Y_t^{M,N,(1)} &= -\lambda \int_0^t e^{(t-s)(\triangle - m_0^2)} \\ &\qquad P_{M,N}^* \left\{ (P_{M,N} P_{2N} Y_s^{M,N})^3 - 3C_*^{(M,N)} \rho_M^2 P_{M,N} P_{2N} Y_s^{M,N} \right\} ds\end{aligned}$$

for $t \in [0, \tau_L)$. Since from the definition of $P_{M,N}$

$$\sup_{s \in [0,t]} \left\| (P_{M,N} P_{2N} Y_s^{M,N})^3 - 3C_*^{(M,N)} \rho_M^2 P_{M,N} P_{2N} Y_s^{M,N} \right\|_{B_2^\beta} < \infty$$

for $t \in [0, \tau_L)$ and $\beta \in \mathbb{R}$, we have $P_{2N} Y_t^{M,N,(1)} \in B_2^\beta$ for $t \in [0, \tau_L)$ and $\beta \in \mathbb{R}$. Hence, from these facts we have, for any $t \in [0, T]$,

$$\begin{aligned}
&\sup_{s \in [0, t \wedge \tau_L]} \left\| P_{2N} Y_s^{M,N,(1)} \right\|_{L^2}^2 \\
&= 2 \sup_{s \in [0, t \wedge \tau_L]} \int_0^s \int_{\mathbb{R}^3} P_{2N} Y_u^{M,N,(1)} \partial_t P_{2N} Y_u^{M,N,(1)} dx du \\
&\leq 2 \sup_{s \in [0, t \wedge \tau_L]} \int_0^s \int_{\mathbb{R}^3} (P_{2N} Y_u^{M,N,(1)}) (\triangle - m_0^2) (P_{2N} Y_u^{M,N,(1)}) dx du \\
&\quad + 2\lambda \sup_{s \in [0, t \wedge \tau_L]} \left\{ - \int_0^s \int_{\mathbb{R}^3} (P_{2N} Y_u^{M,N,(1)}) \left(P_{M,N}^* \left[(P_{M,N} P_{2N} Y_u^{M,N})^3 \right] \right) dx du \right\} \\
&\quad + 6\lambda \left\| C_*^{(M,N)} \right\|_{L^\infty} \sup_{s \in [0, t \wedge \tau_L]} \int_0^s \int_{\mathbb{R}^3} |P_{2N} Y_u^{M,N,(1)}| \left| P_{M,N}^* P_{M,N} P_{2N} Y_u^{M,N} \right| dx du \\
&\leq 2 \sup_{s \in [0, t \wedge \tau_L]} \left\{ - \int_0^s \left(m_0^2 \| \nabla P_{2N} Y_u^{M,N,(1)} \|_{L^2}^2 + \| \nabla P_{2N} Y_u^{M,N,(1)} \|_{L^2}^2 \right) du \right\} \\
&\quad + \lambda \sup_{s \in [0, t \wedge \tau_L]} \left\{ - \int_0^s \| P_{M,N} P_{2N} Y_u^{M,N} \|_{L^4}^4 du \right. \\
&\qquad\qquad \left. + \int_0^s \int_{\mathbb{R}^3} (P_{M,N} P_{2N} Z_u^Y) (P_{M,N} P_{2N} Y_u^{M,N})^3 dx du \right\} \\
&\quad + 6\lambda \left\| C_*^{(M,N)} \right\|_{L^\infty} \sup_{s \in [0, t \wedge \tau_L]} \int_0^s \int_{\mathbb{R}^3} |P_{2N} Y_u^{M,N,(1)}| \left| P_{M,N}^* P_{M,N} P_{2N} Z_u^Y \right| dx du \\
&\quad + 6\lambda \left\| C_*^{(M,N)} \right\|_{L^\infty} \\
&\qquad \times \sup_{s \in [0, t \wedge \tau_L]} \int_0^s \int_{\mathbb{R}^3} |P_{2N} Y_u^{M,N,(1)}| \left| P_{M,N}^* P_{M,N} P_{2N} Y_u^{M,N,(1)} \right| dx du.
\end{aligned}$$

By Hölder's inequality we obtain

$$\begin{aligned}
&\sup_{s\in[0,t\wedge\tau_L]} \left\|P_{2N}Y_s^{M,N,(1)}\right\|_{L^2}^2 \\
&\leq C_{M,N}\left(\sup_{s\in[0,T]}\left\|P_{M,N}Z_u^Y\right\|_{L^4}^4+\sup_{s\in[0,T]}\left\|P_{M,N}Z_u^Y\right\|_{L^2}^2\right.\\
&\qquad\left.+\int_0^t \sup_{u\in[0,s\wedge\tau_L]}\left\|P_{2N}Y_u^{M,N,(1)}\right\|_{L^2}^2 ds\right)
\end{aligned}$$

where $C_{M,N}$ is a constant depending on M and N. Similarly to Proposition 3.3 we have

$$E\left[\sup_{s\in[0,T]}\left\|P_{M,N}Z_u^Y\right\|_{L^4}^4\right]+E\left[\sup_{s\in[0,T]}\left\|P_{M,N}Z_u^Y\right\|_{L^2}^2\right]<\infty$$

for each $M,N\in\mathbb{N}$. Hence,

$$\begin{aligned}
&E\left[\sup_{s\in[0,t\wedge\tau_L]}\left\|P_{2N}Y_s^{M,N,(1)}\right\|_{L^2}^2\right]\\
&\leq C_{M,N}\left(1+\int_0^t E\left[\sup_{u\in[0,s\wedge\tau_L]}\left\|P_{2N}Y_u^{M,N,(1)}\right\|_{L^2}^2\right]ds\right),
\end{aligned}$$

and by applying Gronwall's inequality we get

$$E\left[\sup_{s\in[0,t\wedge\tau_L]}\left\|P_{2N}Y_s^{M,N,(1)}\right\|_{L^2}^2\right]\leq \tilde{C}_{M,N}.$$

where $\tilde{C}_{M,N}$ is a constant depending on M, N and T. This inequality implies

$$E\left[\sup_{s\in[0,T]}\left\|P_{2N}Y_s^{M,N,(1)}\right\|_{L^2(\nu)}^2\right]\leq \lim_{L\to\infty}E\left[\sup_{s\in[0,T\wedge\tau_L]}\left\|P_{2N}Y_s^{M,N,(1)}\right\|_{L^2}^2\right]\leq C_{M,N}.$$

Similarly to Proposition 3.3, by noting that $Z_0^Y=Y_0^{M,N}\in B_2^{-\alpha}(\nu)$, we also have

$$E\left[\sup_{s\in[0,T]}\left\|Z_u^Y\right\|_{B_2^{-\alpha-\varepsilon}(\nu)}^2\right]+E\left[\sup_{s\in[0,T]}\left\|P_{2N}^{\perp}Y_s^{M,N}\right\|_{B_2^{-\alpha-\varepsilon}(\nu)}^2\right]<\infty.$$

Therefore, $Y_t^{M,N}=P_{2N}^{\perp}Y_t^{M,N}+P_{2N}Y_t^{M,N,(1)}+P_{2N}Z_t^Y$ is a time global solution to (4.2) as a continuous process in $B_2^{-\alpha-\varepsilon}(\nu)$. Moreover, the following estimate holds:

$$E\left[\sup_{s\in[0,T]}\left\|Y_s^{M,N}\right\|_{B_2^{-\alpha-\varepsilon}(\nu)}^2\right]<\infty.$$

Since the uniqueness locally in time and the global existence of the solution are obtained for all initial value $Y^{M,N}\in B_2^{-\alpha}(\nu)$ and $\alpha\in(1/2,\infty)$, we have the uniqueness globally in time and the Markov property by a standard argument of the theory of stochastic partial differential equations. For the invariance of $\mu_{M,N}$ under the solution of (4.2), see [**AR91**] for the details. □

Similarly to the argument after the proof of [**AK20**, Theorem 4.1], we are able to construct a pair of random variables $(\xi_{M,N},\zeta)$ whose law is invariant for the system $(Y_t^{M,N}, Z_t)$. We fix a pair of random variables $(\xi_{M,N}, \zeta)$ and consider the stochastic partial differential equation

$$\begin{cases} \partial_t X_t^{M,N}(x) &= \dot{W}_t(x) - (-\triangle + m_0^2) X_t^{M,N}(x) \\ &\quad -\lambda P_{M,N}^* \Big\{ (P_{M,N} X_t^{M,N})^3(x) \\ &\qquad -3C_*^{(M,N)}(x) \rho_M(x)^2 P_{M,N} X_t^{M,N}(x) \Big\} \\ X_0^{M,N}(x) &= \xi_{M,N}(x) \end{cases} \tag{4.6}$$

where $\dot{W}_t$ is a Gaussian white noise independent of $(\xi_{M,N}, \zeta)$. Since the law of ζ is the stationary measure for the solution to (3.2) and independent of $\{\dot{W}_t; t \in [0, \infty)\}$, we are able to choose $\{\dot{W}_t; t \in (-\infty, 0)\}$ so that $\{\dot{W}_t; t \in (-\infty, \infty)\}$ is a Gaussian white noise and that $\zeta = \int_{-\infty}^0 e^{-s(\triangle - m_0^2)} dW_s$. We remark that (4.6) is the same as (1.11) in Section 1.2. Define Z_t by (3.2) with the same $\dot{W}$ in (4.6) and with $Z_0 = \zeta$, where ζ is the random variable defined above. We remark that the pair $(X_t^{M,N}, Z_t)$ is a stationary process for $t \in [0, \infty)$ by the construction of $(\xi_{M,N}, \zeta)$.

Now we transform the SPDE (4.6) similarly as we did for a corresponding SPDE in [**AK20**]. Let $X_t^{M,N,(1)} := X_t^{M,N} - Z_t$ for $t \in [0, \infty)$. From (4.6) and (3.2) we have

$$\begin{aligned} &\partial_t X_t^{M,N,(1)} + (-\triangle + m_0^2) X_t^{M,N,(1)} \\ &= -\lambda P_{M,N}^* \left[(P_{M,N} X_t^{M,N,(1)})^3 \right] dt - 3\lambda P_{M,N}^* \left[\mathcal{Z}_t^{(1,M,N)} (P_{M,N} X_t^{M,N,(1)})^2 \right] \\ &\quad -3\lambda P_{M,N}^* \left[\mathcal{Z}_t^{(2,M,N)} P_{M,N} X_t^{M,N,(1)} \right] dt - \lambda P_{M,N}^* \mathcal{Z}_t^{(3,M,N)} \\ &\quad -9\lambda^2 P_{M,N}^* \left[C_2^{(M,N)} \rho_M^2 \left(P_{M,N} X_t^{M,N,(1)} + \mathcal{Z}_t^{(1,M,N)} \right) \right]. \end{aligned} \tag{4.7}$$

By letting $P_{2N}^\perp := I - P_{2N}$, we have

$$\partial_t P_{2N}^\perp (X_t^{M,N} - Z_t) = (\triangle - m_0^2) P_{2N}^\perp (X_t^{M,N} - Z_t),$$

because $P_{2N}^\perp P_N = 0$. Hence, by the stationarity of $(X_t^{M,N}, Z_t)$ we get

$$\begin{aligned} &E\left[\| P_{2N}^\perp (\xi_{M,N} - \zeta) \|_{B_2^{-1/2-\varepsilon}(\nu)}^2 \right] \\ &= E\left[\| P_{2N}^\perp (X_t^{M,N} - Z_t) \|_{B_2^{-1/2-\varepsilon}(\nu)}^2 \right] \\ &\leq E\left[\| e^{t(\triangle - m_0^2)} P_{2N}^\perp (\xi_{M,N} - \zeta) \|_{B_2^{-1/2-\varepsilon}(\nu)}^2 \right] \\ &\leq e^{-2m_0^2 t} E\left[\| P_{2N}^\perp (\xi_{M,N} - \zeta) \|_{B_2^{-1/2-\varepsilon}(\nu)}^2 \right] \end{aligned}$$

for $t \in [0, \infty)$ and $\varepsilon \in (0, \infty)$. Hence, by letting $t \to \infty$ we have

$$E\left[\| P_{2N}^\perp (\xi_{M,N} - \zeta) \|_{B_2^{-1/2-\varepsilon}(\nu)}^2 \right] = 0.$$

This implies that $P_{2N}^\perp (\xi_{M,N} - \zeta) = 0$ almost surely in $B_2^{-1/2-\varepsilon}(\nu)$. Hence, $\Delta_j (\xi_{M,N} - \zeta) = 0$ except for finitely many $j \in \mathbb{N} \cup \{-1, 0\}$. Thus, we have

$$X_0^{M,N,(1)} = \xi_{M,N} - \zeta \in \bigcap_{s \in \mathbb{R}} B_2^s(\nu) \quad \text{almost surely} \tag{4.8}$$

for each $M, N \in \mathbb{N}$. Let

$$X_t^{M,N,(2)} := X_t^{M,N} - Z_t + \lambda \mathcal{Z}_t^{(0,3,M,N)}, \quad t \in [0, \infty).$$

Then, we have

$$
\begin{aligned}
&\partial_t X_t^{M,N,(2)} + (-\triangle + m_0^2) X_t^{M,N,(2)} \\
&= -\lambda P_{M,N}^* \left[\left(P_{M,N} X_t^{M,N,(2)} - \lambda P_{M,N} \mathcal{Z}_t^{(0,3,M,N)} \right)^3 \right] \\
&\quad -3\lambda P_{M,N}^* \left[\mathcal{Z}_t^{(1,M,N)} \left(P_{M,N} X_t^{M,N,(2)} - \lambda P_{M,N} \mathcal{Z}_t^{(0,3,M,N)} \right)^2 \right] \\
&\quad -3\lambda P_{M,N}^* \left[\mathcal{Z}_t^{(2,M,N)} \left(P_{M,N} X_t^{M,N,(2)} - \lambda P_{M,N} \mathcal{Z}_t^{(0,3,M,N)} \right) \right] \\
&\quad -9\lambda^2 P_{M,N}^* \Big[C_2^{(M,N)} \rho_M^2 \\
&\qquad \times \left(P_{M,N} X_t^{M,N,(2)} + \mathcal{Z}_t^{(1,M,N)} - \lambda P_{M,N} \mathcal{Z}_t^{(0,3,M,N)} \right) \Big].
\end{aligned} \tag{4.9}
$$

Hence, the pair $(X_t^{M,N,(2),<}, X_t^{M,N,(2),\geqslant})$ defined by

$$
\begin{aligned}
X_t^{M,N,(2),<} &:= -3\lambda \int_0^t e^{(t-s)(\triangle - m_0^2)} \\
&\qquad P_{M,N}^* \left[\left(P_{M,N} X_s^{M,N,(2)} - \lambda P_{M,N} \mathcal{Z}_s^{(0,3,M,N)} \right) \circledless \mathcal{Z}_s^{(2,M,N)} \right] ds \\
X_t^{M,N,(2),\geqslant} &:= X_t^{M,N,(2)} - X_t^{M,N,(2),<}
\end{aligned}
$$

is the solution to the following partial differential equation

$$
\begin{cases}
(\partial_t - \triangle + m_0^2) X_t^{M,N,(2),<} \\
= -3\lambda P_{M,N}^* \Big[\Big(P_{M,N} X_t^{M,N,(2),<} + P_{M,N} X_t^{M,N,(2),\geqslant} \\
\qquad\qquad -\lambda P_{M,N} \mathcal{Z}_t^{(0,3,M,N)} \Big) \circledless \mathcal{Z}_t^{(2,M,N)} \Big] \\
\\
(\partial_t - \triangle + m_0^2) X_t^{M,N,(2),\geqslant} \\
= -\lambda P_{M,N}^* \Big[\Big(P_{M,N} X_t^{M,N,(2),<} + P_{M,N} X_t^{M,N,(2),\geqslant} \\
\qquad\qquad -\lambda P_{M,N} \mathcal{Z}_t^{(0,3,M,N)} \Big)^3 \Big] \\
\quad -3\lambda P_{M,N}^* \Big[\mathcal{Z}_t^{(1,N)} \Big(P_{M,N} X_t^{M,N,(2),<} + P_{M,N} X_t^{M,N,(2),\geqslant} \\
\qquad\qquad -\lambda P_{M,N} \mathcal{Z}_t^{(0,3,M,N)} \Big)^2 \Big] \\
\quad -3\lambda P_{M,N}^* \Big[\Big(P_{M,N} X_t^{M,N,(2),<} + P_{M,N} X_t^{M,N,(2),\geqslant} \\
\qquad\qquad -\lambda P_{M,N} \mathcal{Z}_t^{(0,3,M,N)} \Big) \circledgeq \mathcal{Z}_t^{(2,M,N)} \Big] \\
\quad -9\lambda^2 P_{M,N}^* \Big[C_2^{(M,N)} \rho_M^2 \Big(P_{M,N} X_t^{M,N,(2),<} + P_{M,N} X_t^{M,N,(2),\geqslant} \\
\qquad\qquad + \mathcal{Z}_t^{(1,M,N)} - \lambda P_{M,N} \mathcal{Z}_t^{(0,3,M,N)} \Big) \Big]
\end{cases} \tag{4.10}
$$

with initial condition $(X_0^{M,N,(2),<}, X_0^{M,N,(2),\geqslant}) = (0, X_0^{M,N,(2)})$. Now, we change (4.10) for another equivalent equation by using the calculus of paraproducts. By denoting

$$\begin{aligned}
&\Psi_t^{(1,M,N)}(w)\\
&:= P_{M,N}\int_0^t e^{(t-s)(\triangle-m_0^2)}P_{M,N}^*\left[\left(w_s-\lambda P_{M,N}\mathcal{Z}_s^{(0,3,M,N)}\right)\circledless\mathcal{Z}_s^{(2,M,N)}\right]ds\\
&\qquad-\rho_M^2\left[\left(w_t-\lambda P_{M,N}\mathcal{Z}_t^{(0,3,M,N)}\right)\circledless\int_0^t e^{(t-s)(\triangle-m_0^2)}P_N^2\mathcal{Z}_s^{(2,M,N)}ds\right],\\
&\Psi_t^{(2,M,N)}(w)\\
&:= \left(\rho_M^2\left[\left(w_t-\lambda P_{M,N}\mathcal{Z}_t^{(0,3,M,N)}\right)\circledless\int_0^t e^{(t-s)(\triangle-m_0^2)}P_N^2\mathcal{Z}_s^{(2,M,N)}ds\right]\right)\\
&\qquad\qquad\qquad\qquad\qquad\qquad\qquad\qquad\circleddash\mathcal{Z}_t^{(2,M,N)}\\
&\qquad-\rho_M^2\left(w_t-\lambda P_{M,N}\mathcal{Z}_t^{(0,3,M,N)}\right)\\
&\qquad\qquad\qquad\times\left[\int_0^t e^{(t-s)(\triangle-m_0^2)}P_N^2\mathcal{Z}_s^{(2,M,N)}ds\circleddash\mathcal{Z}_t^{(2,M,N)}\right]
\end{aligned}$$

for $w \in C([0,\infty);\mathcal{S}'(\mathbb{R}^3))$, we have

$$\begin{aligned}
&(P_{M,N}X_t^{M,N,(2),<})\circleddash\mathcal{Z}_t^{(2,M,N)}\\
&= -3\lambda\Bigg(P_{M,N}\int_0^t e^{(t-s)(\triangle-m_0^2)}\\
&\qquad P_{M,N}^*\left[\left(P_{M,N}X_s^{M,N,(2)}-\lambda P_{M,N}\mathcal{Z}_s^{(0,3,M,N)}\right)\circledless\mathcal{Z}_s^{(2,M,N)}\right]ds\Bigg)\circleddash\mathcal{Z}_t^{(2,M,N)}\\
&= -3\lambda\rho_M^2\left(P_{M,N}X_t^{M,N,(2)}-\lambda P_{M,N}\mathcal{Z}_t^{(0,3,M,N)}\right)\\
&\qquad\qquad\times\left[\int_0^t e^{(t-s)(\triangle-m_0^2)}P_N^2\mathcal{Z}_s^{(2,M,N)}ds\circleddash\mathcal{Z}_t^{(2,M,N)}\right]\\
&\quad-3\lambda\Psi_t^{(1,M,N)}(P_{M,N}X^{M,N,(2)})\circleddash\mathcal{Z}_t^{(2,M,N)}-3\lambda\Psi_t^{(2,M,N)}(P_{M,N}X^{M,N,(2)})
\end{aligned}$$

for $t \in [0,\infty)$. In view of this equality, (4.10) is equivalent to

$$
\left\{
\begin{aligned}
&(\partial_t - \triangle + m_0^2) X_t^{M,N,(2),<} \\
&= -3\lambda P_{M,N}^* \Big[\Big(P_{M,N} X_t^{M,N,(2),<} + P_{M,N} X_t^{M,N,(2),\geqslant} \\
&\qquad\qquad - \lambda P_{M,N} \mathcal{Z}_t^{(0,3,M,N)} \Big) \varolessthan \mathcal{Z}_t^{(2,M,N)} \Big] \\
&(\partial_t - \triangle + m_0^2) X_t^{M,N,(2),\geqslant} \\
&= -\lambda P_{M,N}^* \Big[\Big(P_{M,N} X_t^{M,N,(2),<} + P_{M,N} X_t^{M,N,(2),\geqslant} \Big)^3 \Big] \\
&\quad + \lambda P_{M,N}^* \Phi_t^{(1,M,N)} (P_{M,N} X_t^{M,N,(2),<} + P_{M,N} X_t^{M,N,(2),\geqslant}) \\
&\quad + \lambda P_{M,N}^* \Phi_t^{(2,M,N)} (P_{M,N} X_t^{M,N,(2),<} + P_{M,N} X_t^{M,N,(2),\geqslant}) \\
&\quad + \lambda P_{M,N}^* \Phi_t^{(3,M,N)} (P_{M,N} X_t^{M,N,(2),<} + P_{M,N} X_t^{M,N,(2),\geqslant}) \\
&\quad - 3\lambda P_{M,N}^* \Big[(P_{M,N} X_t^{M,N,(2),\geqslant}) \circleddash \mathcal{Z}_t^{(2,M,N)} \Big] \\
&\quad + 9\lambda^2 P_{M,N}^* \Big[\Psi_t^{(1,M,N)} (P_{M,N} X^{M,N,(2),<} + P_{M,N} X^{M,N,(2),\geqslant}) \\
&\qquad\qquad \circleddash \mathcal{Z}_t^{(2,M,N)} \Big] \\
&\quad + 9\lambda^2 P_{M,N}^* \Psi_t^{(2,M,N)} (P_{M,N} X^{M,N,(2),<} + P_{M,N} X^{M,N,(2),\geqslant}),
\end{aligned}
\right.
\tag{4.11}
$$

where for $w \in C([0,\infty); \mathcal{S}'(\mathbb{R}^3))$

$$
\begin{aligned}
\Phi_t^{(1,M,N)}(w) &:= -3 \left(\mathcal{Z}_t^{(1,M,N)} - \lambda P_{M,N} \mathcal{Z}_t^{(0,3,M,N)} \right) \circledleqslant w_t^2 \\
&\quad + 3\lambda \left[\left(2\mathcal{Z}_t^{(1,M,N)} - \lambda P_{M,N} \mathcal{Z}_t^{(0,3,M,N)} \right) P_{M,N} \mathcal{Z}_t^{(0,3,M,N)} \right] \circledleqslant w_t, \\
\Phi_t^{(2,M,N)}(w) &:= -3 \left(w_t - \lambda P_{M,N} \mathcal{Z}_t^{(0,3,M,N)} \right) \varogreaterthan \mathcal{Z}_t^{(2,M,N)} + 3\lambda \widehat{\mathcal{Z}}_t^{(2,3,M,N)} \\
&\quad + 9\lambda \left(w_t - \lambda P_{M,N} \mathcal{Z}_t^{(0,3,M,N)} \right) \\
&\qquad \times \left(\mathcal{Z}_t^{(2,2,M,N)} - \mathcal{Z}_t^{(2,M,N)} \circleddash \int_{-\infty}^0 e^{(t-s)(\triangle - m_0^2)} P_N^2 \mathcal{Z}_s^{(2,M,N)} ds \right) \\
&\quad - \lambda^2 \left(3\mathcal{Z}_t^{(1,M,N)} - P_{M,N} \mathcal{Z}_t^{(0,3,M,N)} \right) \left(P_{M,N} \mathcal{Z}_t^{(0,3,M,N)} \right)^2, \\
\Phi_t^{(3,M,N)}(w) &:= -3 \left(\mathcal{Z}_t^{(1,M,N)} - \lambda P_{M,N} \mathcal{Z}_t^{(0,3,M,N)} \right) \varogreaterthan w_t^2 \\
&\quad + 3\lambda \left[\left(2\mathcal{Z}_t^{(1,M,N)} - \lambda P_{M,N} \mathcal{Z}_t^{(0,3,M,N)} \right) P_{M,N} \mathcal{Z}_t^{(0,3,M,N)} \right] \varogreaterthan w_t.
\end{aligned}
$$

To show the tightness of the laws of $\{X^{M,N}; M, N \in \mathbb{N}\}$, we consider uniform estimates similarly to those in [**AK20**]. For $\eta \in [0,1)$, $\gamma \in (0,1/4)$ and $\varepsilon \in (0,1]$

we define $\mathfrak{X}^{M,N}_{\lambda,\alpha,\eta,\gamma}(t)$ and $\mathfrak{Y}^{M,N}_{\varepsilon}(t)$ by

$$\begin{aligned}
&\mathfrak{X}^{M,N}_{\lambda,\alpha,\eta,\gamma}(t)\\
&:= \int_0^t \left(m_0^2 \left\| X_s^{M,N,(2)} \right\|^2_{L^2(\nu)} + \left\| \nabla X_s^{M,N,(2),\geqslant} \right\|^2_{L^2(\nu)} + \lambda \left\| P_{M,N} X_s^{M,N,(2)} \right\|^4_{L^4(\nu)} \right) ds\\
&\quad + \sup_{s',t' \in [0,t]; s'<t'} \frac{(s')^\eta \left\| X_{t'}^{M,N,(2)} - X_{s'}^{M,N,(2)} \right\|_{B^\alpha_{4/3}(\nu)}}{(t'-s')^\gamma},\\
&\mathfrak{Y}^{M,N}_{\varepsilon}(t)\\
&:= \int_0^t \left\| X_s^{M,N,(2),<} \right\|^3_{B^{1-(\varepsilon/12)}_{4-\varepsilon}(\nu)} ds + \int_0^t \left\| X_s^{M,N,(2),\geqslant} \right\|_{B^{1+\varepsilon}_{(4/3)+\varepsilon}(\nu)} ds
\end{aligned}$$

respectively. We remark that $\mathfrak{X}^{M,N}_{\lambda,\alpha,\eta,\gamma}(t)$ takes larger by α than that in [**AK20**]. This improvement can be achieved in a similar way to the proofs of [**AK20**] (see [**Kus20**]). We are going to estimate $E\left[\mathfrak{X}^{M,N}_{\lambda,\alpha,\eta,\gamma}(T)\right]$ and $E\left[\mathfrak{Y}^{M,N}_{\varepsilon}(T)^q\right]$ for given $T \in (0,\infty)$ and $q \in (1, 8/7)$. However, since we construct the estimates in weighted Besov spaces, we have a serious problem which does not appear in [**AK20**]. The problem is that the approximation operator P_N for smoothing is not a symmetric operator on $L^2(\nu)$. In order to overcome this problem, we shall prepare some estimates about the asymptotics of $X^{M,N}$ in N for fixed M in next section.

To simplify the notation, we denote by Q a positive polynomial built with the following quantities

$$\begin{aligned}
&2^{-\beta N} \sup_{t\in[0,T]} \left\| \mathcal{Z}_t^{(1,M,N)} \right\|_{B_p^{\beta-(1/2)-\varepsilon}(\nu)}, \quad 2^{-\beta N} \sup_{t\in[0,T]} \left\| \mathcal{Z}_t^{(2,M,N)} \right\|_{B_p^{\beta-1-\varepsilon}(\nu)},\\
&2^{-\beta N} \sup_{t\in[0,T]} \left\| \mathcal{Z}_t^{(0,2,M,N)} \right\|_{B_p^{\beta+1-\varepsilon}(\nu)}, \quad 2^{-\beta N} \sup_{t\in[0,T]} \left\| \mathcal{Z}_t^{(0,3,M,N)} \right\|_{B_p^{\beta+(1/2)-\varepsilon}(\nu)},\\
&2^{-\beta N} \sup_{t\in[0,T]} \left\| \mathcal{Z}_t^{(2,2,M,N)} \right\|_{B_p^{\beta-\varepsilon}(\nu)}, \quad 2^{-\beta N} \sup_{t\in[0,T]} \left\| \widehat{\mathcal{Z}}_t^{(2,3,M,N)} \right\|_{B_p^{\beta-(1/2)-\varepsilon}(\nu)},\\
&2^{-\beta N} \sup_{t\in[0,T]} \left\| \mathcal{Z}_t^{(1,M,N)} \left(P_{M,N} \mathcal{Z}_t^{(0,3,N)} \right) \right\|_{B_p^{\beta-(1/2)-\varepsilon}(\nu)},\\
&2^{-\beta N} \sup_{t\in[0,T]} \left\| \mathcal{Z}_t^{(1,M,N)} \left(P_{M,N} \mathcal{Z}_t^{(0,3,N)} \right)^2 \right\|_{B_p^{\beta-(1/2)-\varepsilon}(\nu)}\\
&\text{and} \quad \sup_{\substack{s,t\in[0,T]\\ s<t}} \frac{\left\| \mathcal{Z}_t^{(0,3,M,N)} - \mathcal{Z}_s^{(0,3,M,N)} \right\|_{L^p(\nu)}}{(t-s)^\gamma},
\end{aligned}$$

for $\beta \in [0,\infty)$, $p \in [1,\infty)$ and $\varepsilon \in (0,1]$ with coefficients depending on λ_0, α, ε, η, γ and T, and we also denote by C a positive constant depending on λ_0, ε, η, γ and T. Positive constants depending on an extra parameter δ are denoted by C_δ. We remark that Q, C and C_δ can be different from line to line and that Proposition 3.3 implies $E[Q] \leq C$ for some C. Moreover, in view of Lemma 2.11 constants of the form $C(1 + M^\kappa 2^{-\delta N})$ appear many times in the estimate in the following sections. So, for simplicity we denote $C(1 + M^\kappa 2^{-\delta N})$ for some $\kappa \in \mathbb{N}$ and $\delta \in (0,1)$ by $K_{M,N}$.

CHAPTER 5

Bounds for the behaviour of $X^{M,N,(2)}$ in N

In this section we consider asymptotics of $X^{M,N,(2)}$ in N for fixed M, which appear in commutator estimates between the smoothing operator P_N and the weight ν. Besides, we often consider Besov norms without weights. As in Proposition 2.4 we need change of weights for the Besov embedding theorem. However, for fixed M by using the localization of interactions by ρ_M we are able to obtain sufficiently good asymptotics of Besov norms without weights.

We use the same notation as in Chapter 4. In this section, for simplicity we denote

$$\begin{aligned}
F_t^{(1)}(\mathcal{Z}) &= \mathcal{Z}_t^{(1,M,N)} - \lambda P_{M,N}\mathcal{Z}_t^{(0,3,M,N)}\\
F_t^{(2)}(\mathcal{Z}) &= \mathcal{Z}_t^{(2,M,N)} - 2\lambda\mathcal{Z}_t^{(1,M,N)}P_{M,N}\mathcal{Z}_t^{(0,3,M,N)} + (\lambda P_{M,N}\mathcal{Z}_t^{(0,3,M,N)})^2\\
F_t^{(3)}(\mathcal{Z}) &= 3\lambda\Big(\widehat{\mathcal{Z}}_t^{(2,3,M,N)} + \mathcal{Z}_t^{(2,M,N)}\circledless P_{M,N}\mathcal{Z}_t^{(0,3,M,N)}\\
&\qquad\qquad\qquad\qquad\qquad + \mathcal{Z}_t^{(2,M,N)}\circledgtr P_{M,N}\mathcal{Z}_t^{(0,3,M,N)}\Big)\\
&\qquad - 3\mathcal{Z}_t^{(1,M,N)}(\lambda P_{M,N}\mathcal{Z}_t^{(0,3,M,N)})^2 + (\lambda P_{M,N}\mathcal{Z}_t^{(0,3,M,N)})^3.
\end{aligned}$$

We remark that $F_t^{(1)}(\mathcal{Z})$, $F_t^{(2)}(\mathcal{Z})$ and $F_t^{(3)}(\mathcal{Z})$ depend on M and N.

LEMMA 5.1. *Let $\varepsilon \in (0,1)$. For each $p \in [1,\infty)$ it holds that*

$$\begin{aligned}
\sup_{t\in[0,T]} \|F_t^{(1)}(\mathcal{Z})\|_{B_p^{\alpha-\varepsilon}(\nu)} &\le Q2^{(\alpha+1/2)N}, \quad \alpha > -\frac{1}{2}\\
\sup_{t\in[0,T]} \|F_t^{(2)}(\mathcal{Z})\|_{B_p^{\alpha-\varepsilon}(\nu)} &\le Q2^{(\alpha+1)N}, \quad \alpha > -1\\
\sup_{t\in[0,T]} \|F_t^{(3)}(\mathcal{Z})\|_{B_p^{\alpha-\varepsilon}(\nu)} &\le Q2^{(\alpha+1)N}, \quad \alpha > -1.
\end{aligned}$$

PROOF. The estimates follows from Propositions 2.5 and 3.3. □

PROPOSITION 5.2. For any $\varepsilon \in (0,1)$ and $M, N \in \mathbb{N}$ it holds that

$$\int_0^T E\left[\|X_s^{M,N,(2)}\|_{L^2}^2 + \|\nabla X_s^{M,N,(2)}\|_{L^2}^2 + \lambda\|P_{M,N}X_s^{M,N,(2)}\|_{L^4}^4\right] ds \le K_{M,N}2^{\varepsilon N}.$$

Moreover, for each $t \in [0,T]$, it holds that

$$\begin{aligned}
\|X_t^{M,N,(2)}\|_{L^2}^2 + \int_0^t \left(\|X_s^{M,N,(2)}\|_{L^2}^2 + \|\nabla X_s^{M,N,(2)}\|_{L^2}^2\right) ds&\\
+ \lambda\int_0^t \|P_{M,N}X_s^{M,N,(2)}\|_{L^4}^4 ds&\\
\le C\|X_0^{M,N,(2)}\|_{L^2}^2 + K_{M,N}2^{\varepsilon N}Q < \infty&
\end{aligned}$$

for $M, N \in \mathbb{N}$ almost surely.

PROOF. Let $h_n(x) := \exp(n^{-1}\sqrt{1+|x|^2})$ for $x \in \mathbb{R}^3$ and $n \in \mathbb{N}$. From (4.9) we have

$$
\begin{aligned}
&\|X_t^{M,N,(2)}\|_{L^2(h_n)}^2 - \|X_0^{M,N,(2)}\|_{L^2(h_n)}^2 \\
&= 2\int_0^t \int_{\mathbb{R}^3} X_s^{M,N,(2)}[(\triangle - m_0^2)X_s^{M,N,(2)}]h_n dxds \\
&\quad - 2\lambda \int_0^t \int_{\mathbb{R}^3} X_s^{M,N,(2)} \left(P_{M,N}^*\left[(P_{M,N}X_s^{M,N,(2)})^3\right]\right) h_n dxds \\
&\quad - 6\lambda \int_0^t \int_{\mathbb{R}^3} X_s^{M,N,(2)} \left(P_{M,N}^*\left[F_s^{(1)}(\mathcal{Z})(P_{M,N}X_s^{M,N,(2)})^2\right]\right) h_n dxds \\
&\quad - 6\lambda \int_0^t \int_{\mathbb{R}^3} X_s^{M,N,(2)} \left(P_{M,N}^*\left[F_s^{(2)}(\mathcal{Z})P_{M,N}X_s^{M,N,(2)}\right]\right) h_n dxds \\
&\quad + 2\lambda \int_0^t \int_{\mathbb{R}^3} X_s^{M,N,(2)} \left(P_{M,N}^* F_s^{(3)}(\mathcal{Z})\right) h_n dxds \\
&\quad + 18\lambda^2 \int_0^t \int_{\mathbb{R}^3} X_s^{M,N,(2)} \\
&\qquad\quad \times P_{M,N}^*\left[C_2^{(M,N)}\rho_M^2\left(P_{M,N}X_s^{M,N,(2)} - \lambda P_{M,N}\mathcal{Z}_t^{(0,3,M,N)}\right)\right] h_n dxds \\
&= -2\int_0^t \left(\|\nabla X_s^{M,N,(2)}\|_{L^2(h_n)}^2 + m_0^2\|X_s^{M,N,(2)}\|_{L^2(h_n)}^2\right) ds \\
&\quad - 2\int_0^t \int_{\mathbb{R}^3} X_s^{M,N,(2)}\nabla X_s^{M,N,(2)} \cdot \nabla h_n dxds - 2\lambda \int_0^t \int_{\mathbb{R}^3} (P_{M,N}X_s^{M,N,(2)})^4 dxds \\
&\quad + 2\lambda \int_0^t \int_{\mathbb{R}^3} (1-h_n)X_s^{M,N,(2)} P_{M,N}^*\left[(P_{M,N}X_s^{M,N,(2)})^3\right] dxds \\
&\quad - 6\lambda \int_0^t \int_{\mathbb{R}^3} F_s^{(1)}(\mathcal{Z})\nu^{-1}(P_{M,N}X_s^{M,N,(2)})^3 h_n \nu dxds \\
&\quad + 6\lambda \int_0^t \int_{\mathbb{R}^3} (1-h_n)X_s^{M,N,(2)} \left(P_{M,N}^*\left[F_s^{(1)}(\mathcal{Z})(P_{M,N}X_s^{M,N,(2)})^2\right]\right) dxds \\
&\quad - 6\lambda \int_0^t \int_{\mathbb{R}^3} F_s^{(2)}(\mathcal{Z})\nu^{-1}(P_{M,N}X_s^{M,N,(2)})^2 h_n \nu dxds \\
&\quad + 6\lambda \int_0^t \int_{\mathbb{R}^3} (1-h_n)X_s^{M,N,(2)} \left(P_{M,N}^*\left[F_s^{(2)}(\mathcal{Z})P_{M,N}X_s^{M,N,(2)}\right]\right) dxds \\
&\quad + 2\lambda \int_0^t \int_{\mathbb{R}^3} F_s^{(3)}(\mathcal{Z})\nu^{-1}(P_{M,N}X_s^{M,N,(2)}) h_n \nu dxds \\
&\quad - 2\lambda \int_0^t \int_{\mathbb{R}^3} (1-h_n)X_s^{M,N,(2)} P_{M,N}^* F_s^{(3)}(\mathcal{Z}) dxds \\
&\quad + 18\lambda^2 \int_0^t \int_{\mathbb{R}^3} C_2^{(M,N)}\nu^{-1}\rho_M^2 \\
&\qquad\quad \times \left[P_{M,N}(h_n X_s^{M,N,(2)})\right]\left[P_{M,N}X_s^{N,(2)} - \lambda P_{M,N}\mathcal{Z}_s^{(0,3,M,N)}\right] \nu dxds.
\end{aligned}
$$

Hence, Proposition 2.4 and Lemma 2.9 imply

$$
\begin{aligned}
&\|X_t^{M,N,(2)}\|_{L^2(h_n)}^2 - \|X_0^{M,N,(2)}\|_{L^2(h_n)}^2 \\
&\le -2\int_0^t \left(\|\nabla X_s^{M,N,(2)}\|_{L^2(h_n)}^2 + m_0^2\|X_s^{M,N,(2)}\|_{L^2(h_n)}^2\right) ds \\
&\quad - 2\int_0^t \int_{\mathbb{R}^3} X_s^{M,N,(2)} \nabla X_s^{M,N,(2)} \cdot \nabla h_n dx ds - 2\lambda \int_0^t \int_{\mathbb{R}^3} (P_{M,N} X_s^{M,N,(2)})^4 dx ds \\
&\quad + 2\lambda \int_0^t \int_{\mathbb{R}^3} (1-h_n) X_s^{M,N,(2)} P_{M,N}^* \left[(P_{M,N} X_s^{M,N,(2)})^3\right] dx ds \\
&\quad + \lambda C \left(\sup_{s\in[0,T]} \left\|F_s^{(1)}(\mathcal{Z})\right\|_{B_{6/\varepsilon}^{-1/2+\varepsilon}(\nu)}\right) \int_0^t \left\|\nu^{-1}(P_{M,N} X_s^{M,N,(2)})^3\right\|_{B_{6/(6-\varepsilon)}^{(1-\varepsilon)/2}(\nu)} ds \\
&\quad + 6\lambda \int_0^t \int_{\mathbb{R}^3} (1-h_n) X_s^{M,N,(2)} \left(P_{M,N}^* \left[F_s^{(1)}(\mathcal{Z})(P_{M,N} X_s^{M,N,(2)})^2\right]\right) dx ds \\
&\quad + \lambda C \left(\sup_{s\in[0,T]} \left\|F_s^{(2)}(\mathcal{Z})\right\|_{B_{6/\varepsilon}^{-1+\varepsilon}(\nu)}\right) \int_0^t \left\|\nu^{-1}(P_{M,N} X_s^{M,N,(2)})^2\right\|_{B_{6/(6-\varepsilon)}^{1-\varepsilon/2}(\nu)} ds \\
&\quad + 6\lambda \int_0^t \int_{\mathbb{R}^3} (1-h_n) X_s^{M,N,(2)} \left(P_{M,N}^* \left[F_s^{(2)}(\mathcal{Z}) P_{M,N} X_s^{M,N,(2)}\right]\right) dx ds \\
&\quad + \lambda C \left(\sup_{s\in[0,T]} \left\|F_s^{(3)}(\mathcal{Z})\right\|_{B_2^{-1}(\nu)}\right) \int_0^t \|\nu^{-1} P_{M,N} X_s^{M,N,(2)}\|_{B_2^1(\nu)} ds \\
&\quad + 2\lambda \int_0^t \int_{\mathbb{R}^3} (1-h_n) X_s^{M,N,(2)} \left(P_{M,N}^* F_s^{(3)}(\mathcal{Z})\right) dx ds \\
&\quad + 18\lambda^2 \int_0^t \left\|C_2^{(M,N)} \nu^{-1} \rho_M^2 P_{M,N}\left(h_n X_s^{M,N,(2)}\right)\right\|_{L^2} \left\|P_{M,N} X_s^{M,N,(2)}\right\|_{L^2} ds \\
&\quad + 18\lambda^3 \left(\sup_{t\in[0,T]} \left\|P_{M,N} \mathcal{Z}_t^{(0,3,M,N)}\right\|_{L^2(\nu)}\right) \\
&\qquad\qquad \times \int_0^t \left\|C_2^{(M,N)} \nu^{-1} \rho_M^2 P_{M,N}\left(h_n X_s^{M,N,(2)}\right)\right\|_{L^2(\nu)} ds.
\end{aligned}
$$

Note that from Propositions 2.2 and 2.4, Lemma 2.7 and the fact that ν^{-1} and the derivative of ν^{-1} are bounded by CM^σ, it follows that

$$
\begin{aligned}
&\left\|\nu^{-1}(P_{M,N} X_s^{M,N,(2)})^3\right\|_{B_{6/(6-\varepsilon)}^{(1-\varepsilon)/2}(\nu)} \\
&\le \left\|\nu^{-1}(P_{M,N} X_s^{M,N,(2)})^3\right\|_{B_{6/(6-\varepsilon)}^{(1-\varepsilon)/2}} \\
&\le C \left\|\nu^{-1}(P_{M,N} X_s^{M,N,(2)})^3\right\|_{B_1^{1/2}} \\
&\le C \left\|\nu^{-1}(P_{M,N} X_s^{M,N,(2)})^3\right\|_{L^1}^{1/2} \left\|\nu^{-1}(P_{M,N} X_s^{M,N,(2)})^3\right\|_{W^{1,1}}^{1/2} \\
&\le \delta \left\|(P_{M,N} X_s^{M,N,(2)})^3\right\|_{W^{1,1}(\nu)} + CM^{2\sigma}\delta^{-1} \left\|(P_{M,N} X_s^{M,N,(2)})^3\right\|_{L^1(\nu)} \\
&\le \delta \left(\|P_{M,N} X_s^{M,N,(2)}\|_{L^4(\nu)}^4 + \|\nabla P_{M,N} X_s^{M,N,(2)}\|_{L^2(\nu)}^2\right) + CM^{14\sigma/3}\delta^{-7/3}
\end{aligned}
$$

for any $\delta \in (0,1]$, and similarly it follows that

$$\begin{aligned}&\left\|\nu^{-1}(P_{M,N}X_s^{M,N,(2)})^2\right\|_{B^{1-\varepsilon/2}_{6/(6-\varepsilon)}(\nu)}\\ &\le \delta\left\|\nabla P_{M,N}X_s^{M,N,(2)}\right\|^2_{L^2(\nu)} + \delta\left\|P_{M,N}X_s^{M,N,(2)}\right\|^4_{L^4(\nu)} + CM^{6\sigma}\delta^{-3}\end{aligned}$$

for any $\delta \in (0,1]$. Moreover, in view of Proposition 3.2

$$\begin{aligned}&\left\|C_2^{(M,N)}\nu^{-1}\rho_M^2 P_{M,N}\left(h_n X_s^{M,N,(2)}\right)\right\|_{L^2(\nu)}\\ &\le CM^\sigma N\|X_s^{M,N,(2)}\|_{L^2(\nu)}\\ &\le \delta\|X_s^{M,N,(2)}\|^2_{L^2(\nu)} + CM^{2\sigma}N^2\delta^{-1}.\end{aligned}$$

Hence, by applying Lemma 2.7 and Proposition 3.2, for $\delta \in (0,1]$ we have

$$\begin{aligned}&\|X_t^{M,N,(2)}\|^2_{L^2(h_n)} - \|X_0^{M,N,(2)}\|^2_{L^2(h_n)}\\ &+2\int_0^t\left(\|\nabla X_s^{M,N,(2)}\|^2_{L^2(h_n)} + m_0^2\|X_s^{M,N,(2)}\|^2_{L^2(h_n)}\right)ds\\ &+2\lambda\int_0^t\left\|P_{M,N}X_s^{M,N,(2)}\right\|^4_{L^4}ds\\ &\le -2\int_0^t\int_{\mathbb{R}^3}X_s^{M,N,(2)}\nabla X_s^{M,N,(2)}\cdot\nabla h_n dxds\\ &\quad+\lambda\delta\int_0^t\left(\left\|\nabla X_s^{M,N,(2)}\right\|^2_{L^2} + \left\|X_s^{M,N,(2)}\right\|^2_{L^2} + \left\|P_{M,N}X_s^{M,N,(2)}\right\|^4_{L^4}\right)ds\\ &\quad+2\lambda\int_0^t\int_{\mathbb{R}^3}(1-h_n)X_s^{M,N,(2)}P^*_{M,N}\left[(P_{M,N}X_s^{M,N,(2)})^3\right]dxds\\ &\quad+6\lambda\int_0^t\int_{\mathbb{R}^3}(1-h_n)X_s^{M,N,(2)}\left(P^*_{M,N}\left[F_s^{(1)}(\mathcal{Z})(P_{M,N}X_s^{M,N,(2)})^2\right]\right)dxds\\ &\quad+6\lambda\int_0^t\int_{\mathbb{R}^3}(1-h_n)X_s^{M,N,(2)}\left(P^*_{M,N}\left[F_s^{(2)}(\mathcal{Z})P_{M,N}X_s^{M,N,(2)}\right]\right)dxds\\ &\quad+2\lambda\int_0^t\int_{\mathbb{R}^3}(1-h_n)X_s^{M,N,(2)}\left(P^*_{M,N}F_s^{(3)}(\mathcal{Z})\right)dxds + C_\delta M^{6\sigma}2^{6\varepsilon N}Q.\end{aligned}\tag{5.1}$$

Note that the stationarity of $X^{N,(1)}$ implies

$$\begin{aligned}&E\left[\|X_t^{M,N,(2)}\|^2_{L^2(h_n)}\right] - E\left[\|X_0^{M,N,(2)}\|^2_{L^2(h_n)}\right]\\ &= 2\lambda E\left[\int_{\mathbb{R}^3}\mathcal{Z}_t^{(0,3,M,N)}X_t^{M,N,(2)}h_n dx\right] - \lambda^2 E\left[\|\mathcal{Z}_t^{(0,3,M,N)}\|^2_{L^2(h_n)}\right]\\ &\quad - 2\lambda E\left[\int_{\mathbb{R}^3}\mathcal{Z}_0^{(0,3,M,N)}X_0^{M,N,(2)}h_n dx\right] + \lambda^2 E\left[\|\mathcal{Z}_0^{(0,3,M,N)}\|^2_{L^2(h_n)}\right],\end{aligned}$$

and that Hölder's inequality, Lemma 2.11, Proposition 3.3, and the stationarity of $X^{M,N,(1)}$ imply

$$\begin{aligned}
&\sup_{s\in[0,t]} E\left[\left|\int_{\mathbb{R}^3} \mathcal{Z}_s^{(0,3,M,N)} X_s^{M,N,(2)} h_n dx\right|\right] \\
&\le \delta \sup_{s\in[0,t]} E\left[\|X_s^{M,N,(2)}\|_{L^2(h_n)}^2\right] + C_\delta K_{M,N} \sup_{s\in[0,t]} E\left[\|\mathcal{Z}_s^{(0,3,M,N)}\|_{L^2(h_n)}^2\right] \\
&\le 2\delta \sup_{s\in[0,t]} E\left[\|X_s^{M,N,(1)}\|_{L^2(h_n)}^2\right] \\
&\quad + 2\lambda^2\delta \sup_{s\in[0,t]} E\left[\|\mathcal{Z}_s^{(0,3,M,N)}\|_{L^2(h_n)}^2\right] + C_\delta K_{M,N} M^\sigma \\
&\le \frac{2\delta}{T} E\left[\int_0^T \|X_s^{M,N,(1)}\|_{L^2(h_n)}^2 ds\right] + C_\delta K_{M,N} M^\sigma \\
&\le \frac{2\delta}{T} E\left[\int_0^T \|X_s^{M,N,(2)}\|_{L^2(h_n)}^2 ds\right] + C_\delta K_{M,N} M^\sigma
\end{aligned}$$

for $\delta \in (0,1]$. Since $|\nabla h_n| \le (2n)^{-1} h_n$ (recalling that $h_n \ge 0$), by taking expectation first and then limit $n \to \infty$ in (5.1), we have

$$\begin{aligned}
&\int_0^T E\left[\|X_s^{M,N,(2)}\|_{L^2}^2 + \|\nabla X_s^{M,N,(2)}\|_{L^2}^2\right] ds + \lambda \int_0^T E\left[\|P_{M,N} X_s^{M,N,(2)}\|_{L^4}^4\right] ds \\
&\le \lambda C\delta \int_0^T E\left[\left\|X_s^{M,N,(2)}\right\|_{L^2}^2 + \left\|\nabla X_s^{M,N,(2)}\right\|_{L^2}^2 + \left\|P_{M,N} X_s^{M,N,(2)}\right\|_{L^4}^4\right] ds \\
&\quad + \frac{C\delta}{T} E\left[\int_0^T \|X_s^{M,N,(2)}\|_{L^2}^2 ds\right] + C_\delta K_{M,N} M^{6\sigma} 2^{6\varepsilon N}
\end{aligned}$$

for $\delta \in (0,1]$. Since $\varepsilon \in (0,1)$ is arbitrary, by taking δ sufficiently small, we get the first assertion of the lemma.

The first assertion implies in particular that

$$\int_0^t \|\nabla X_s^{M,N,(2)}\|_{L^2}^2 ds, \qquad \int_0^t \|P_{M,N} X_s^{M,N,(2)}\|_{L^4}^4 ds$$

are finite almost surely for each $t \in [0,T]$. Moreover, the first assertion and the stationarity of $X^{M,N,(1)}$, Lemma 2.11 and Proposition 3.3 also imply that

$$\begin{aligned}
&E\left[\|X_t^{M,N,(2)}\|_{L^2}^2\right] \\
&\le 2E\left[\|X_t^{M,N,(1)}\|_{L^2}^2\right] + 2\lambda E\left[\|\mathcal{Z}_t^{(0,3,M,N)}\|_{L^2}^2\right] \\
&\le \frac{2}{T} E\left[\int_0^T \|X_s^{M,N,(1)}\|_{L^2}^2 ds\right] + CM^\sigma \\
&\le \frac{4}{T} E\left[\int_0^T \|X_s^{M,N,(2)}\|_{L^2}^2 ds\right] + \frac{4}{T} E\left[\int_0^T \|\mathcal{Z}_t^{(0,3,M,N)}\|_{L^2}^2 ds\right] + CM^\sigma \\
&< \infty.
\end{aligned}$$

Hence, $\|X_t^{M,N,(2)}\|_{L^2}^2$ is also finite almost surely for each $t \in [0,T]$. In view of these facts, by letting $n \to \infty$ in (5.1) we have

$$\begin{aligned}
&\|X_t^{M,N,(2)}\|_{L^2}^2 - \|X_0^{M,N,(2)}\|_{L^2}^2 \\
&+ 2\int_0^t \left(\|\nabla X_s^{M,N,(2)}\|_{L^2}^2 + m_0^2\|X_s^{M,N,(2)}\|_{L^2}^2\right) ds + 2\lambda \int_0^t \left\|P_{M,N}X_s^{M,N,(2)}\right\|_{L^4}^4 ds \\
&\leq \lambda\delta \int_0^t \left(\left\|X_s^{M,N,(2)}\right\|_{L^2}^2 + \left\|\nabla X_s^{M,N,(2)}\right\|_{L^2}^2 + \left\|P_{M,N}X_s^{M,N,(2)}\right\|_{L^4}^4\right) ds \\
&\quad + C_\delta M^{6\sigma} 2^{6\varepsilon N} Q.
\end{aligned}$$

almost surely for $\delta \in (0,1]$ and $t \in [0,T]$. By taking δ sufficiently small we get the second assertion of the lemma. □

COROLLARY 5.3. For $\alpha \in (1/2, \infty)$ and $M, N \in \mathbb{N}$

$$2^{-\alpha N} \sup_{t\in[0,T]} E\left[\|X_t^{M,N,(2)}\|_{B_2^{\frac{1}{2}}}^2\right] \leq K_{M,N}.$$

PROOF. For any $\varepsilon \in (0,1)$, by the stationarity of $X^{M,N,(1)}$, Lemma 2.11 and Proposition 3.3 we have

$$\begin{aligned}
&\sup_{t\in[0,T]} E\left[\|X_t^{M,N,(2)}\|_{B_2^{\frac{1}{2}}}^2\right] \\
&\leq 2 \sup_{t\in[0,T]} E\left[\|X_t^{M,N,(1)}\|_{B_2^{\frac{1}{2}}}^2\right] + 2\lambda^2 \sup_{t\in[0,T]} E\left[\|\mathcal{Z}_t^{(0,3,M,N)}\|_{B_2^{\frac{1}{2}}}^2\right] \\
&\leq \frac{2}{T}\int_0^T E\left[\|X_t^{M,N,(1)}\|_{B_2^{\frac{1}{2}}}^2\right] dt + CM^\sigma 2^{(1+\varepsilon)N/2} \\
&\leq \frac{2}{T}\int_0^T E\left[\|X_t^{M,N,(2)}\|_{B_2^{\frac{1}{2}}}^2\right] dt + CM^\sigma 2^{(1+\varepsilon)N/2}.
\end{aligned}$$

Hence, by Proposition 5.2 we obtain the assertion. □

PROPOSITION 5.4. For $\alpha \in [0, 1/2]$, sufficiently small $\varepsilon \in (0,1)$ and $t \in [0,T]$

$$\begin{aligned}
\|X_t^{M,N,(2)}\|_{B_2^\alpha} \leq C\|X_0^{M,N,(2)}\|_{B_2^\alpha} &+ K_{M,N}2^{[(3+2\varepsilon)/4]N}\|X_0^{M,N,(2)}\|_{L^2}^{3/2} \\
&+ K_{M,N}2^{(1+\varepsilon)N}\|X_0^{M,N,(2)}\|_{L^2}^{1/2} + K_{M,N}2^{[(9/8)+\varepsilon]N}Q.
\end{aligned}$$

PROOF. In view of the mild form of the solution to (4.9) we have

$$
\begin{aligned}
&X_t^{M,N,(2)} \\
&= e^{t(\triangle - m_0^2)} X_0^{M,N,(2)} - \lambda \int_0^t e^{(t-s)(\triangle - m_0^2)} P_{M,N}^* \left[(P_{M,N} X_s^{M,N,(2)})^3 \right] ds \\
&\quad - 3\lambda \int_0^t e^{(t-s)(\triangle - m_0^2)} P_{M,N}^* \left[F_s^{(1)}(\mathcal{Z}) (P_{M,N} X_s^{M,N,(2)})^2 \right] ds \\
&\quad - 3\lambda \int_0^t e^{(t-s)(\triangle - m_0^2)} P_{M,N}^* \left[F_s^{(2)}(\mathcal{Z}) P_{M,N} X_s^{M,N,(2)} \right] ds \\
&\quad + \lambda \int_0^t e^{(t-s)(\triangle - m_0^2)} P_{M,N}^* F_s^{(3)}(\mathcal{Z}) ds \\
&\quad + 9\lambda^2 \int_0^t e^{(t-s)(\triangle - m_0^2)} \\
&\qquad\qquad P_{M,N}^* \left[C_2^{(M,N)} \rho_M^2 \left(P_{M,N} X_s^{M,N,(2)} - \lambda P_{M,N} \mathcal{Z}_s^{(0,3,M,N)} \right) \right] ds
\end{aligned}
$$

for $t \in [0,T]$. Hence, by Lemmas 2.9 and 2.11 and Proposition 2.2 we have

$$
\begin{aligned}
&\|X_t^{M,N,(2)}\|_{B_2^\alpha} \\
&\leq C \|X_0^{M,N,(2)}\|_{B_2^\alpha} \\
&\quad + \lambda C \int_0^t (t-s)^{-(1-\varepsilon)/4} \left\| P_{M,N}^* \left[(P_{M,N} X_s^{M,N,(2)})^3 \right] \right\|_{B_2^{\alpha-(1-\varepsilon)/2}} ds \\
&\quad + \lambda C \int_0^t (t-s)^{-(1-\varepsilon)/2} \left\| P_{M,N}^* \left[F_s^{(1)}(\mathcal{Z}) (P_{M,N} X_s^{M,N,(2)})^2 \right] \right\|_{B_2^{\alpha-1+\varepsilon}} ds \\
&\quad + \lambda C \int_0^t (t-s)^{-\alpha/2} \left\| P_{M,N}^* \left[F_s^{(2)}(\mathcal{Z}) P_{M,N} X_s^{M,N,(2)} \right] \right\|_{B_2^0} ds \\
&\quad + \lambda C \int_0^t (t-s)^{-\alpha/2} \left\| P_{M,N}^* F_s^{(3)}(\mathcal{Z}) \right\|_{B_2^0} ds \\
&\quad + \lambda^2 C \int_0^t (t-s)^{-\alpha/2} \left\| P_{M,N}^* \left[C_2^{(M,N)} \rho_M^2 P_{M,N} X_s^{M,N,(2)} \right] \right\|_{B_2^0} ds \\
&\quad + \lambda^2 C \int_0^t (t-s)^{-\alpha/2} \left\| P_{M,N}^* \left[C_2^{(M,N)} \rho_M^2 P_{M,N} \mathcal{Z}_s^{(0,3,M,N)} \right] \right\|_{B_2^0} ds \\
&\leq C \|X_0^{N,(2)}\|_{B_2^\alpha} \\
&\quad + \lambda K_{M,N} 2^{[\alpha+(1+2\varepsilon)/4]N} \int_0^t (t-s)^{-(1-\varepsilon)/4} \left\| (P_{M,N} X_s^{M,N,(2)})^3 \right\|_{B_2^{-3/4}} ds \\
&\quad + \lambda K_{M,N} 2^{[(1/4)+\varepsilon]N} \int_0^t (t-s)^{-(1-\varepsilon)/2} \left\| F_s^{(1)}(\mathcal{Z}) (P_{M,N} X_s^{M,N,(2)})^2 \right\|_{B_2^{-3/4}} ds \\
&\quad + \lambda K_{M,N} 2^{\varepsilon N} \int_0^t (t-s)^{-\alpha/2} \left\| F_s^{(2)}(\mathcal{Z}) P_{M,N} X_s^{M,N,(2)} \right\|_{B_2^{-\varepsilon}} ds \\
&\quad + \lambda K_{M,N} \int_0^t (t-s)^{-\alpha/2} \left\| \rho_M F_s^{(3)}(\mathcal{Z}) \right\|_{L^2} ds \\
&\quad + \lambda^2 K_{M,N} \int_0^t (t-s)^{-\alpha/2} \left\| C_2^{(M,N)} \rho_M^2 P_{M,N} X_s^{M,N,(2)} \right\|_{L^2} ds
\end{aligned}
$$

$$+ \lambda^2 K_{M,N} \int_0^t (t-s)^{-\alpha/2} \left\| C_2^{(M,N)} \rho_M^2 P_{M,N} \mathcal{Z}_s^{(0,3,M,N)} \right\|_{L^2} ds.$$

Hence, from Propositions 2.2, 2.4, 2.5 and 3.2, it follows that

$$\begin{aligned}
&\|X_t^{M,N,(2)}\|_{B_2^\alpha} \\
&\leq C\|X_0^{M,N,(2)}\|_{B_2^\alpha} \\
&\quad + \lambda K_{M,N} 2^{[\alpha+(1+2\varepsilon)/4]N} \int_0^t (t-s)^{-(1-\varepsilon)/4} \left\| (P_{M,N} X_s^{M,N,(2)})^3 \right\|_{B^0_{4/3}} ds \\
&\quad + \lambda K_{M,N} 2^{[(1/4)+\varepsilon]N} \int_0^t (t-s)^{-(1-\varepsilon)/2} \left\| F_s^{(1)}(\mathcal{Z}) (P_{M,N} X_s^{M,N,(2)})^2 \right\|_{B^0_{4/3}} ds \\
&\quad + \lambda K_{M,N} 2^{\varepsilon N} \int_0^t (t-s)^{-\alpha/2} \left\| F_s^{(2)}(\mathcal{Z}) P_{M,N} X_s^{M,N,(2)} \right\|_{B^0_{6/(3+2\varepsilon)}} ds \\
&\quad + \lambda K_{M,N} \int_0^t (t-s)^{-\alpha/2} \left\| \rho_M F_s^{(3)}(\mathcal{Z}) \right\|_{L^2} ds \\
&\quad + \lambda^2 K_{M,N} N \int_0^t (t-s)^{-\alpha/2} \left\| \rho_M^2 P_{M,N} X_s^{M,N,(2)} \right\|_{L^2} ds \\
&\quad + \lambda^2 K_{M,N} N \sup_{s\in[0,T]} \left\| \rho_M^2 P_{M,N} \mathcal{Z}_s^{(0,3,M,N)} \right\|_{L^2} \\
&\leq C\|X_0^{M,N,(2)}\|_{B_2^\alpha} \\
&\quad + \lambda K_{M,N} M^\sigma 2^{[\alpha+(1+2\varepsilon)/4]N} \int_0^t (t-s)^{-(1-\varepsilon)/4} \left\| (P_{M,N} X_s^{M,N,(2)})^3 \right\|_{L^{4/3}(\nu)} ds \\
&\quad + \lambda K_{M,N} M^\sigma 2^{[(1/4)+\varepsilon]N} \int_0^t (t-s)^{-(1-\varepsilon)/2} \left\| F_s^{(1)}(\mathcal{Z}) (P_{M,N} X_s^{M,N,(2)})^2 \right\|_{L^{4/3}(\nu)} ds \\
&\quad + \lambda K_{M,N} M^\sigma 2^{\varepsilon N} \int_0^t (t-s)^{-\alpha/2} \left\| F_s^{(2)}(\mathcal{Z}) P_{M,N} X_s^{M,N,(2)} \right\|_{L^{6/(3+2\varepsilon)}(\nu)} ds \\
&\quad + \lambda K_{M,N} M^\sigma \int_0^t (t-s)^{-\alpha/2} \left\| F_s^{(3)}(\mathcal{Z}) \right\|_{L^2(\nu)} ds \\
&\quad + \lambda^2 K_{M,N} M^\sigma N \int_0^t (t-s)^{-\alpha/2} \left\| \rho_M^2 P_{M,N} X_s^{M,N,(2)} \right\|_{L^2(\nu)} ds \\
&\quad + K_{M,N} M^\sigma N \sup_{s\in[0,T]} \left\| \mathcal{Z}_s^{(0,3,M,N)} \right\|_{L^2(\nu)} \\
&\leq C\|X_0^{M,N,(2)}\|_{B_2^\alpha} \\
&\quad + \lambda K_{M,N} M^\sigma 2^{[\alpha+(1+2\varepsilon)/4]N} \int_0^t (t-s)^{-(1-\varepsilon)/4} \left\| P_{M,N} X_s^{M,N,(2)} \right\|_{L^4(\nu)}^3 ds \\
&\quad + \lambda K_{M,N} M^\sigma 2^{[(1/4)+\varepsilon]N} \left(\sup_{s\in[0,T]} \left\| F_s^{(1)}(\mathcal{Z}) \right\|_{L^4(\nu)} \right) \\
&\qquad\qquad \times \int_0^t (t-s)^{-(1-\varepsilon)/2} \left\| (P_{M,N} X_s^{M,N,(2)})^2 \right\|_{L^2(\nu)} ds \\
&\quad + \lambda K_{M,N} M^\sigma 2^{\varepsilon N} \left(\sup_{s\in[0,T]} \left\| F_s^{(2)}(\mathcal{Z}) \right\|_{L^{3/\varepsilon}(\nu)} \right)
\end{aligned}$$

$$\times \int_0^t (t-s)^{-\alpha/2} \left\| P_{M,N} X_s^{M,N,(2)} \right\|_{L^2(\nu)} ds$$
$$+ \lambda^2 K_{M,N} M^\sigma N \int_0^t (t-s)^{-\alpha/2} \left\| P_{M,N} X_s^{M,N,(2)} \right\|_{L^2(\nu)} ds$$
$$+ \lambda K_{M,N} M^\sigma \sup_{s\in[0,T]} \left\| F_s^{(3)}(\mathcal{Z}) \right\|_{L^2(\nu)} + K_{M,N} M^\sigma N \sup_{s\in[0,T]} \left\| \mathcal{Z}_s^{(0,3,M,N)} \right\|_{L^2(\nu)}$$

for $t \in [0,T]$. From this inequality, Lemma 2.7 and Propositions 2.2 and 5.2 we have

$$\begin{aligned}
&\|X_t^{M,N,(2)}\|_{B_2^\alpha} \\
&\leq C\|X_0^{M,N,(2)}\|_{B_2^\alpha} + \lambda K_{M,N} M^\sigma 2^{[\alpha+(1+2\varepsilon)/4]N} \left(\int_0^t \left\| P_{M,N} X_s^{M,N,(2)} \right\|_{L^4(\nu)}^4 ds \right)^{3/4} \\
&\quad + \lambda K_{M,N} M^\sigma 2^{[(3/4)+\varepsilon]N} \left(\int_0^t \|P_{M,N} X_s^{M,N,(2)}\|_{L^4(\nu)}^4 ds \right)^{1/2} \\
&\quad + \lambda K_{M,N} M^\sigma 2^{(1+\varepsilon)N} \left(\int_0^t \|P_{M,N} X_s^{M,N,(2)}\|_{L^4(\nu)}^4 ds \right)^{1/4} + \lambda K_{M,N} M^\sigma 2^{(1+\varepsilon)N} Q \\
&\leq C\|X_0^{M,N,(2)}\|_{B_2^\alpha} + K_{M,N} M^\sigma 2^{[(3+2\varepsilon)/4]N} \|X_0^{M,N,(2)}\|_{L^2}^{3/2} \\
&\quad + K_{M,N} M^\sigma 2^{(1+\varepsilon)N} \|X_0^{M,N,(2)}\|_{L^2}^{1/2} + K_{M,N} M^\sigma 2^{[(9/8)+2\varepsilon]N} Q.
\end{aligned}$$

Since $\varepsilon \in (0,1)$ is arbitrary, this is the desired inequality. □

COROLLARY 5.5. For $\alpha \in [0,1/2]$ and $\beta \in (9/8,\infty)$

$$2^{-(4\beta/3)N} E\left[\sup_{s\in[0,T]} \|X_s^{M,N,(2)}\|_{B_2^\alpha}^{4/3} \right] \leq K_{M,N}.$$

PROOF. Let $\varepsilon \in (0,1)$ sufficiently small. Proposition 5.4 implies that

$$\begin{aligned}
&E\left[\sup_{s\in[0,T]} \|X_t^{M,N,(2)}\|_{B_2^\alpha}^{4/3} \right]^{3/4} \\
&\leq CE\left[\|X_0^{M,N,(2)}\|_{B_2^\alpha}^{4/3} \right]^{3/4} + K_{M,N} 2^{[(3/4)+\varepsilon]N} E\left[\|X_0^{M,N,(2)}\|_{L^2}^2 \right]^{3/4} \\
&\quad + K_{M,N} 2^{(1+\varepsilon)N} E\left[\|X_0^{M,N,(2)}\|_{L^2}^{2/3} \right]^{3/4} + K_{M,N} 2^{[(9/8)+\varepsilon]N} \\
&\leq CE\left[\|X_0^{M,N,(2)}\|_{B_2^\alpha}^2 \right]^{1/2} + K_{M,N} 2^{[(3/4)+\varepsilon]N} E\left[\|X_0^{M,N,(2)}\|_{L^2}^2 \right]^{3/4} \\
&\quad + K_{M,N} 2^{(1+\varepsilon)N} E\left[\|X_0^{M,N,(2)}\|_{L^2}^2 \right]^{1/4} + K_{M,N} 2^{[(9/8)+\varepsilon]N}.
\end{aligned}$$

Hence, by Corollary 5.3 we obtain the assertion. □

CHAPTER 6

Tightness of the laws of $\{X^{M,N}\}$

Now we prove the tightness of the laws of $\{X^{M,N}; M, N \in \mathbb{N}\}$ in the space of continuous stochastic processes on suitable Besov spaces. We choose $\sigma \in (3, \infty)$ which satisfies (4.1), and prove some uniform estimates in the approximation sequence. The argument is similar to that in [**AK20**, Section 4]. However, some estimates are improved from those in [**AK20**] (see [**Kus20**]).

Similarly to the proofs of [**AK20**, Lemmas 4.2, 4.3, 4.4, and 4.6–4.9], we obtain the following proposition.

PROPOSITION 6.1. For $p \in [1, 2]$, $\varepsilon \in (0, 1/16]$ and $\delta \in (0, 1]$, the following estimates hold:

$$\begin{aligned}
&\int_0^t \|X_s^{M,N,(2),<}\|^3_{B^{1-\varepsilon}_{4-\varepsilon}(\nu)} dt \leq \delta \left(\int_0^t \left\| P_{M,N} X_s^{M,N,(2)} \right\|^4_{L^4(\nu)} ds \right)^{7/8} + Q\delta^{-6}, \\
&\left\| \Psi_t^{(2,M,N)}(P_{M,N} X^{M,N,(2)}) \right\|_{B^{\varepsilon}_p(\nu)} \\
&\leq \delta \left(\|P_{M,N} X_t^{M,N,(2)}\|^4_{L^4(\nu)} + \|X_t^{M,N,(2)}\|^2_{B_2^{15/16}(\nu)} \right)^{7/8} + Q\delta^{-16/19}, \\
&\left\| \Phi_t^{(1,M,N)}(P_{M,N} X^{M,N,(2)}) \right\|_{L^{4/3}(\nu)} \\
&\leq \delta \left(\|P_{M,N} X_t^{M,N,(2)}\|^4_{L^4(\nu)} + \|P_{M,N} X_t^{M,N,(2)}\|^2_{B_2^{15/16}(\nu)} \right)^{7/8} + \delta^{-26/9} Q, \\
&\left\| \Phi_t^{(2,M,N)}(P_{M,N} X^{M,N,(2)}) \right\|_{B_p^{-1/2-\varepsilon}(\nu)} \\
&\leq \delta \left(\|P_{M,N} X_t^{M,N,(2)}\|^4_{L^4(\nu)} + \|P_{M,N} X_t^{M,N,(2)}\|^2_{B_2^{15/16}(\nu)} \right)^{7/8} + \delta^{-16/19} t^{-1/4-\varepsilon} Q, \\
&\left\| \Phi_t^{(3,M,N)}(P_{M,N} X^{M,N,(2)}) \right\|_{B_p^{-(1+\varepsilon)/2}(\nu)} \leq \delta \|P_{M,N} X_t^{M,N,(2)}\|^{7/2}_{L^4(\nu)} + \delta^{-4/3} Q, \\
&\left\| (P_{M,N} X_t^{M,N,(2),\geqslant}) \circleddash \mathcal{Z}_t^{(2,M,N)} \right\|_{B_p^{\varepsilon/8}(\nu)} \\
&\leq \delta \left(\left\| \nabla X_t^{M,N,(2),\geqslant} \right\|^2_{L^2(\nu)} + \left\| P_{M,N} X_t^{M,N,(2)} \right\|^4_{L^4(\nu)} \right)^{7/8} \\
&\qquad + \delta \left\| X_t^{M,N,(2),<} \right\|^{7/4}_{L^{p+\varepsilon}(\nu)} + \delta \left\| X_t^{M,N,(2),\geqslant} \right\|^{5/6}_{B^{1+\varepsilon}_{p+\varepsilon}(\nu)} + \delta^{-82/23} Q.
\end{aligned}$$

Some of the estimates in Proposition 6.1 are different from those in [**AK20**] by ε. The reason is that the Wick polynomials $\mathcal{Z}^{(k,M,N)}$ of the Ornstein-Uhlenbeck processes are B_∞^{-s}-valued continuous processes for suitable $s \in (0, \infty)$ in the case of the torus, but not in the case of the whole space $\mathbb{R}^3$. Hence, in the present paper we need to replace the B_∞^{-s}-norms of $\mathcal{Z}^{(k,M,N)}$ in [**AK20**] by the B_p^{-s}-norms of $\mathcal{Z}^{(k,M,N)}$ with sufficiently large $p \in [1,\infty)$ and weight ν.

PROPOSITION 6.2. For $\gamma \in (0,1/8)$, $\eta \in [0,1)$, $p \in [1,2]$, $\varepsilon \in (0,1/16]$ and $\theta \in (\varepsilon/\gamma, 1/2]$, it holds that

$$\begin{aligned}
&\left\| \Psi_t^{(1,M,N)}(P_{M,N}X^{M,N,(2)}) \ominus \mathcal{Z}_t^{(2,M,N)} \right\|_{B_p^\varepsilon(\nu)} \\
&\leq Q \int_0^t (t-s)^{-21/32} \left\| P_{M,N}X_s^{M,N,(2)} \right\|_{B_{p+\varepsilon}^{15/16}(\nu)} ds \\
&\quad + Q \left(\sup_{s\in[0,t]} \frac{s^\eta \left\| P_{M,N}X_t^{M,N,(2)} - P_{M,N}X_s^{M,N,(2)} \right\|_{L^p(\nu)}}{(t-s)^\gamma} \right)^\theta \\
&\quad\quad \times \Big(\| P_{M,N}X_t^{M,N,(2)} \|_{L^{p+5\varepsilon}(\nu)}^{1-\theta} \\
&\quad\quad\quad + \int_0^t s^{-\eta\theta}(t-s)^{\theta\gamma-1-\varepsilon} \left\| P_{M,N}X_s^{M,N,(2)} \right\|_{L^{p+5\varepsilon}(\nu)}^{1-\theta} ds \Big) \\
&\quad + Q.
\end{aligned}$$

PROOF. Similarly to the proof of [**AK20**, Lemma 4.3], we have

$$\begin{aligned}
&\left\| \Psi_t^{(1,M,N)}(P_{M,N}X^{M,N,(2)}) \right\|_{B_{p+(\varepsilon/2)}^\varepsilon(\nu)} \\
&\leq Q \int_0^t (t-s)^{-21/32} \left\| P_{M,N}X_s^{M,N,(2)} \right\|_{B_{p+\varepsilon}^{15/16}(\nu)} ds + Q \\
&\quad + Q \int_0^t (t-s)^{-1-\varepsilon} \left\| P_{M,N}X_t^{M,N,(2)} - P_{M,N}X_s^{M,N,(2)} \right\|_{L^{p+\varepsilon}(\nu)} ds.
\end{aligned}$$

Once we show

$$\begin{aligned}
&\int_0^t (t-s)^{-1-\varepsilon} \left\| P_{M,N}X_t^{M,N,(2)} - P_{M,N}X_s^{M,N,(2)} \right\|_{L^{p+\varepsilon}(\nu)} ds \\
&\leq \left(\sup_{s\in[0,t]} \frac{s^\eta \left\| P_{M,N}X_t^{M,N,(2)} - P_{M,N}X_s^{M,N,(2)} \right\|_{L^p(\nu)}}{(t-s)^\gamma} \right)^\theta \\
&\quad \times \Big(\| P_{M,N}X_t^{M,N,(2)} \|_{L^{p+5\varepsilon}(\nu)}^{1-\theta} \\
&\quad\quad + \int_0^t s^{-\eta\theta}(t-s)^{\theta\gamma-1-\varepsilon} \left\| P_{M,N}X_s^{M,N,(2)} \right\|_{L^{p+5\varepsilon}(\nu)}^{1-\theta} ds \Big),
\end{aligned} \tag{6.1}$$

by following the proof of [**AK20**, Lemma 4.4] we obtain the assertion. Hence, it is sufficient to show the validity of (6.1). The interpolation inequality on weighted L^p-spaces implies

$$\begin{aligned}
&\left\| P_{M,N}X_t^{M,N,(2)} - P_{M,N}X_s^{M,N,(2)} \right\|_{L^{p+\varepsilon}(\nu)} \\
&\leq \left\| P_{M,N}X_t^{M,N,(2)} - P_{M,N}X_s^{M,N,(2)} \right\|_{L^p(\nu)}^{1/2} \left\| P_{M,N}X_t^{M,N,(2)} - P_{M,N}X_s^{M,N,(2)} \right\|_{L^q(\nu)}^{1/2}
\end{aligned}$$

where $q = p(p+\varepsilon)/(p-\varepsilon) \le p + 5\varepsilon$. From this inequality we have

$$
\begin{aligned}
&\int_0^t (t-s)^{-1-\varepsilon} \left\| P_{M,N}X_t^{M,N,(2)} - P_{M,N}X_s^{M,N,(2)} \right\|_{L^{p+\varepsilon}(\nu)} ds \\
&\le \int_0^t (t-s)^{-1-\varepsilon} \left\| P_{M,N}X_t^{M,N,(2)} - P_{M,N}X_s^{M,N,(2)} \right\|_{L^p(\nu)}^{\theta} \\
&\qquad \times \left(\left\| P_{M,N}X_t^{M,N,(2)} \right\|_{L^{p+5\varepsilon}(\nu)} + \left\| P_{M,N}X_s^{M,N,(2)} \right\|_{L^{p+5\varepsilon}(\nu)} \right)^{1-\theta} ds \\
&\le C \left(\sup_{s\in[0,t]} \frac{s^{\eta} \left\| P_{M,N}X_t^{M,N,(2)} - P_{M,N}X_s^{M,N,(2)} \right\|_{L^p(\nu)}}{(t-s)^{\gamma}} \right)^{\theta} \\
&\quad \times \left(\| P_{M,N}X_t^{M,N,(2)} \|_{L^{p+5\varepsilon}(\nu)}^{1-\theta} \right. \\
&\qquad \left. + \int_0^t s^{-\eta\theta}(t-s)^{\theta\gamma-1-\varepsilon} \left\| P_{M,N}X_s^{M,N,(2)} \right\|_{L^{p+5\varepsilon}(\nu)}^{1-\theta} ds \right).
\end{aligned}
$$

Thus, we obtain (6.1). □

REMARK 6.3. In the proof of [**AK20**, Proposition 4.13], we apply the proposition associated to Proposition 6.2 with $\theta = 1$. However, since for $p \in [1,2]$

$$
\begin{aligned}
&\int_0^t Q \left(\sup_{s\in[0,u]} \frac{s^{\eta} \left\| P_{M,N}X_u^{M,N,(2)} - P_{M,N}X_s^{M,N,(2)} \right\|_{L^p(\nu)}}{(u-s)^{\gamma}} \right)^{1/2} \\
&\quad \times \left(\| P_{M,N}X_u^{M,N,(2)} \|_{L^{p+5\varepsilon}(\nu)}^{1/2} \right. \\
&\qquad \left. + \int_0^u s^{-\eta/2}(u-s)^{(\gamma/2)-1-\varepsilon} \left\| P_{M,N}X_s^{M,N,(2)} \right\|_{L^{p+5\varepsilon}(\nu)}^{1/2} ds \right) du \\
&\le Q \left(\sup_{s',t'\in[0,t];s'<t'} \frac{(s')^{\eta} \left\| P_{M,N}X_{t'}^{M,N,(2)} - P_{M,N}X_{s'}^{M,N,(2)} \right\|_{L^p(\nu)}}{(t'-s')^{\gamma}} \right)^{1/2} \\
&\quad \times \left(\int_0^t \| P_{M,N}X_u^{M,N,(2)} \|_{L^4(\nu)}^{1/2} du + \int_0^t s^{-\eta/2} \left\| P_{M,N}X_s^{M,N,(2)} \right\|_{L^4(\nu)}^{1/2} ds \right) \\
&\le \delta \sup_{s',t'\in[0,t];s'<t'} \frac{(s')^{\eta} \left\| P_{M,N}X_{t'}^{M,N,(2)} - P_{M,N}X_{s'}^{M,N,(2)} \right\|_{L^p(\nu)}}{(t'-s')^{\gamma}} \\
&\quad + \delta \int_0^t \| P_{M,N}X_u^{M,N,(2)} \|_{L^4(\nu)}^4 du + C_{\delta} Q,
\end{aligned}
$$

a similarly argument as the one used in [**AK20**] works in the proof of Proposition 6.6 below.

PROPOSITION 6.4. For $M, N \in \mathbb{N}$, $t \in [0,T]$ and $\varepsilon \in (0, 1/16]$,

$$
\begin{aligned}
&\left| \int_{\mathbb{R}^3} X_t^{M,N,(2),\geqslant} P_{M,N}^* \left[\left(P_{M,N} X_t^{M,N,(2)} - \lambda P_{M,N} \mathcal{Z}_t^{(0,3,M,N)} \right) \right.\right. \\
&\qquad\qquad\qquad\qquad\qquad\qquad \left.\left. \circledless \mathcal{Z}_t^{(2,M,N)} \right] \nu dx \right| \\
&\le \delta \left(\|\nabla X_t^{M,N,(2),\geqslant}\|_{L^2(\nu)}^2 + \|X_t^{M,N,(2)}\|_{L^2(\nu)}^2 + \|P_{M,N} X_t^{M,N,(2)}\|_{L^4(\nu)}^4 \right) \\
&\quad + \delta \left(\left\| X_t^{M,N,(2),<} \right\|_{B_{4-\varepsilon}^{1-\varepsilon}(\nu)}^3 + \left\| X_t^{M,N,(2),\geqslant} \right\|_{B_{(4/3)+\varepsilon}^{1+\varepsilon}(\nu)} \right) + C_\delta K_{M,N} Q
\end{aligned}
\tag{6.2}
$$

and the terms

$$\left| \int_{\mathbb{R}^3} X_t^{M,N,(2)} P_{M,N}^* \Phi_t^{(1,M,N)} (P_{M,N} X^{M,N,(2)}) \nu dx \right| \tag{6.3}$$

$$\left| \int_{\mathbb{R}^3} X_t^{M,N,(2)} P_{M,N}^* \Phi_t^{(3,M,N)} (P_{M,N} X^{M,N,(2)}) \nu dx \right| \tag{6.4}$$

$$\left| \int_{\mathbb{R}^3} X_t^{M,N,(2)} P_{M,N}^* \left[(P_{M,N} X_t^{M,N,(2),\geqslant}) \circleddash \mathcal{Z}_t^{(2,M,N)} \right] \nu dx \right| \tag{6.5}$$

$$\left| \int_{\mathbb{R}^3} X_t^{M,N,(2)} P_{M,N}^* \Psi_t^{(2,M,N)} (P_{M,N} X^{M,N,(2)}) \nu dx \right| \tag{6.6}$$

are dominated by

$$
\begin{aligned}
&\delta \left(\left\| \nabla X_t^{M,N,(2),\geqslant} \right\|_{L^2(\nu)}^2 + \left\| X_t^{M,N,(2)} \right\|_{L^2(\nu)}^2 + \left\| P_{M,N} X_t^{M,N,(2)} \right\|_{L^4(\nu)}^4 \right) \\
&+ C M^{4\sigma} 2^{-4N} \left\| X_t^{M,N,(2)} \right\|_{L^4(\nu)}^4 + C_\delta K_{M,N} Q.
\end{aligned}
$$

Moreover,

$$
\begin{aligned}
&\left| \int_{\mathbb{R}^3} X_t^{M,N,(2)} P_{M,N}^* \left[\Psi_t^{(1,M,N)} (P_{M,N} X^{M,N,(2)}) \circleddash \mathcal{Z}_t^{(2,M,N)} \right] \nu dx \right| \\
&\le \delta \sup_{s',t' \in [0,t]; s' < t'} \frac{(s')^\eta \left\| P_{M,N} X_{t'}^{M,N,(2)} - P_{M,N} X_{s'}^{M,N,(2)} \right\|_{L^p(\nu)}}{(t'-s')^\gamma} \\
&\quad + \delta \int_0^t \left\| P_{M,N} X_s^{M,N,(2)} \right\|_{L^4(\nu)}^4 ds + C_\delta K_{M,N} Q,
\end{aligned}
\tag{6.7}
$$

and

$$
\begin{aligned}
&\left| \int_{\mathbb{R}^3} X_t^{M,N,(2)} P_{M,N}^* \Phi_t^{(2,M,N)} (P_{M,N} X^{M,N,(2)}) \nu dx \right| \\
&\le \delta \left(\left\| P_{M,N} X_t^{M,N,(2)} \right\|_{L^4(\nu)}^4 + \left\| P_{M,N} X_t^{M,N,(2)} \right\|_{B_2^{15/16}(\nu)}^2 \right)^{7/8} \\
&\qquad\qquad + C_\delta t^{-1/4-\varepsilon} K_{M,N} Q.
\end{aligned}
\tag{6.8}
$$

PROOF. The duality between $B^{1+\varepsilon/8}_{(4/3)+\varepsilon}(\nu)$ and $B^{-1-\varepsilon/8}_{(4+3\varepsilon)/(1+3\varepsilon)}(\nu)$, Lemma 2.11 and Proposition 2.5 imply

$$\begin{aligned}
&\left|\int_{\mathbb{R}^3} X_t^{M,N,(2),\geqslant} P_{M,N}^* \left[\left(P_{M,N} X_t^{M,N,(2)} - \lambda P_{M,N}\mathcal{Z}_t^{(0,3,M,N)}\right) \circledless \mathcal{Z}_t^{(2,M,N)}\right] \nu dx\right| \\
&\leq \left\|X_t^{M,N,(2),\geqslant}\right\|_{B^{1+\varepsilon/8}_{(4/3)+\varepsilon}(\nu)} \\
&\quad \times \left\|P_{M,N}^* \left[\left(P_{M,N} X_t^{M,N,(2)} - \lambda P_{M,N}\mathcal{Z}_t^{(0,3,M,N)}\right) \circledless \mathcal{Z}_t^{(2,M,N)}\right]\right\|_{B^{-1-\varepsilon/8}_{(4+3\varepsilon)/(1+3\varepsilon)}(\nu)} \\
&\leq K_{M,N} Q \left\|X_t^{M,N,(2),\geqslant}\right\|_{B^{1+\varepsilon/8}_{(4/3)+\varepsilon}(\nu)} \left\|P_{M,N} X_t^{M,N,(2)} - \lambda P_{M,N}\mathcal{Z}_t^{(0,3,M,N)}\right\|_{L^4(\nu)} \\
&\leq K_{M,N} Q \left\|X_t^{M,N,(2),\geqslant}\right\|^{3/2}_{B^{1+\varepsilon/8}_{(4/3)+\varepsilon}(\nu)} + \left\|P_{M,N} X_t^{M,N,(2)}\right\|^3_{L^4(\nu)} + Q.
\end{aligned}$$

Similarly to the proof of [**AK20**, Lemma 4.10], by Proposition 2.4 we have

$$\begin{aligned}
&K_{M,N} Q \left\|X_t^{M,N,(2),\geqslant}\right\|^{3/2}_{B^{1+\varepsilon/8}_{(4/3)+\varepsilon}(\nu)} \\
&\leq C_\delta K_{M,N} Q \left\|X_t^{M,N,(2),\geqslant}\right\|^{1/4}_{B^{1+\varepsilon}_{4/3}(\nu)} \left\|X_t^{M,N,(2),\geqslant}\right\|^{5/4}_{B^{1-\varepsilon/20}_{20(4+3\varepsilon)/(60-9\varepsilon)}(\nu)} \\
&\leq \delta \left\|X_t^{M,N,(2),\geqslant}\right\|_{B^{1+\varepsilon}_{4/3}(\nu)} + C_\delta K_{M,N} Q \left\|X_t^{M,N,(2),\geqslant}\right\|^{5/3}_{B^{1-\varepsilon/20}_{2}(\nu)} \\
&\leq \delta \left\|X_t^{M,N,(2),\geqslant}\right\|_{B^{1+\varepsilon}_{4/3}(\nu)} + \delta \left\|X_t^{M,N,(2),\geqslant}\right\|^{2}_{B^{1-\varepsilon/20}_{2}(\nu)} + C_\delta K_{M,N} Q
\end{aligned}$$

for $\delta \in (0,1]$. Moreover,

$$\left\|P_{M,N} X_t^{M,N,(2)}\right\|^3_{L^4(\nu)} \leq \delta \left\|P_{M,N} X_t^{M,N,(2)}\right\|^4_{L^4(\nu)} + C_\delta$$

for $\delta \in (0,1]$. From these inequalities we obtain the estimate for (6.2).

Next we prove the estimate for (6.3). The duality between $P_{M,N}$ and $P_{M,N}^*$ on $L^2(dx)$ implies

$$\begin{aligned}
&\left|\int_{\mathbb{R}^3} X_t^{M,N,(2)} P_{M,N}^* \Phi_t^{(1,M,N)}(P_{M,N} X^{M,N,(2)}) \nu dx\right| \\
&\leq \left|\int_{\mathbb{R}^3} \left(P_{M,N} X_t^{M,N,(2)}\right) \Phi_t^{(1,M,N)}(P_{M,N} X^{M,N,(2)}) \nu dx\right| \\
&\quad + \left|\int_{\mathbb{R}^3} \left(P_{M,N} X_t^{M,N,(2)} - \frac{1}{\nu} P_{M,N}(\nu X_t^{M,N,(2)})\right) \Phi_t^{(1,M,N)}(P_{M,N} X^{M,N,(2)}) \nu dx\right|
\end{aligned}$$

The first term on the right-hand side is estimated, similarly to the proof of [**AK20**, Lemma 4.10], by Proposition 6.1. To estimate the second term on the right-hand side, applying Hölder's inequality and Lemma 2.12, we dominate it by

$$\begin{aligned}
&\left\| P_{M,N}X_t^{M,N,(2)} - \frac{1}{\nu}P_{M,N}(\nu X_t^{M,N,(2)}) \right\|_{L^4(\nu)} \left\| \Phi_t^{(1,M,N)}(P_{M,N}X^{M,N,(2)}) \right\|_{L^{4/3}(\nu)} \\
&\leq \left(CM^\sigma 2^{-N} \left\| X_t^{M,N,(2)} \right\|_{L^2(\nu)} \right) \\
&\quad \times C \left(\left\| \left(\mathcal{Z}_t^{(1,M,N)} - \lambda P_{M,N}\mathcal{Z}_t^{(0,3,M,N)} \right) \circledcirc (P_{M,N}X^{M,N,(2)})^2 \right\|_{L^{4/3}(\nu)} \right. \\
&\qquad + \left\| \left[\left(2\mathcal{Z}_t^{(1,M,N)} - \lambda P_{M,N}\mathcal{Z}_t^{(0,3,M,N)} \right) P_{M,N}\mathcal{Z}_t^{(0,3,M,N)} \right] \right. \\
&\qquad\qquad \left. \left. \circledcirc P_{M,N}X^{M,N,(2)} \right\|_{L^{4/3}(\nu)} \right) \\
&\leq CM^\sigma 2^{-N} Q \left\| X_t^{M,N,(2)} \right\|_{L^2(\nu)} \\
&\qquad \times \left(\left\| (P_{M,N}X^{M,N,(2)})^2 \right\|_{B_{4/3+\varepsilon}^{(1/2)+\varepsilon}(\nu)} + \left\| P_{M,N}X^{M,N,(2)} \right\|_{B_{4/3+\varepsilon}^{(1/2)+\varepsilon}(\nu)} \right).
\end{aligned}$$

Hence, Lemma 2.7 with the replacement of δ by

$$\delta \left(1 + CM^\sigma 2^{-N} Q \| X_t^{M,N,(2)} \|_{L^2(\nu)} \right)^{-1}$$

yields

$$\begin{aligned}
&\left| \int_{\mathbb{R}^3} \left(P_{M,N}X_t^{M,N,(2)} - \frac{1}{\nu}P_{M,N}(\nu X_t^{M,N,(2)}) \right) \Phi_t^{(1,M,N)}(P_{M,N}X^{M,N,(2)}) \nu dx \right| \\
&\leq \delta \left(\left\| P_{M,N}X^{M,N,(2)} \right\|_{L^4(\nu)}^4 + \left\| P_{M,N}X^{M,N,(2)} \right\|_{B_2^{15/16}(\nu)}^2 \right)^{7/8} \\
&\quad + C_\delta Q \left(1 + CM^\sigma 2^{-N} \left\| X_t^{M,N,(2)} \right\|_{L^2(\nu)} \right)^3 \\
&\leq \delta \left(\| \nabla X_t^{M,N,(2),\geqslant} \|_{L^2(\nu)}^2 + \| X_t^{M,N,(2)} \|_{L^2(\nu)}^2 + \| P_{M,N}X_t^{M,N,(2)} \|_{L^4(\nu)}^4 \right) \\
&\quad + CM^{4\sigma} 2^{-4N} \left\| X_t^{M,N,(2)} \right\|_{L^4(\nu)}^4 + C_\delta K_{M,N} Q
\end{aligned}$$

for $\delta \in (0,1]$. Thus we obtain the estimate for (6.3). Proofs of the estimates for (6.4) – (6.8) are obtained in a similar way (see the proof of [**AK20**, Proposition 4.11] for (6.7)). So, we omit details. □

PROPOSITION 6.5. For $\gamma \in (0,1/8)$, $\varepsilon \in (0,\gamma/4)$, $\eta \in [0,1)$ and $t \in [0,T]$

$$\begin{aligned}
&\left\|X_t^{M,N,(2)}\right\|_{L^2(\nu)}^2 - \left\|X_0^{M,N,(2)}\right\|_{L^2(\nu)}^2 \\
&+ \int_0^t \left(m_0^2 \left\|X_s^{M,N,(2)}\right\|_{L^2(\nu)}^2 + \left\|\nabla X_s^{M,N,(2),\geqslant}\right\|_{L^2(\nu)}^2 + \lambda \left\|P_{M,N}X_s^{M,N,(2)}\right\|_{L^4(\nu)}^4 \right) ds \\
&\le C \left\|X_t^{M,N,(2),<}\right\|_{L^2(\nu)}^2 + C\mathfrak{Y}_\varepsilon^{M,N}(t) + CM^{4\sigma}2^{-4N} \int_0^t \left\|X_s^{M,N,(2)}\right\|_{L^4}^4 ds + K_{M,N}Q \\
&\quad + Q\left(\sup_{s',t' \in [0,t]; s'<t'} \frac{(s')^\eta \left\|P_{M,N}X_{t'}^{M,N,(2)} - P_{M,N}X_{s'}^{M,N,(2)}\right\|_{L^{4/3}(\nu)}}{(t'-s')^\gamma} \right)^{4/5}.
\end{aligned}$$

PROOF. The duality between $P_{M,N}$ and $P_{M,N}^*$ on $L^2(dx)$ implies

$$\begin{aligned}
&\int_0^t \int_{\mathbb{R}^3} X_s^{M,N,(2)} P_{M,N}^* \left[\left(P_{M,N}X_s^{M,N,(2)}\right)^3 \right] \nu dx ds \\
&= \int_0^t \int_{\mathbb{R}^3} |P_{M,N}X_s^{M,N,(2)}|^4 \nu dx ds \\
&\quad - \int_0^t \int_{\mathbb{R}^3} \left[P_{M,N}X_s^{M,N,(2)} - \frac{1}{\nu} P_{M,N}(\nu X_s^{M,N,(2)}) \right] (P_{M,N}X_s^{N,(2)})^3 \nu dx ds.
\end{aligned}$$

On the other hand, by Lemma 2.12

$$\begin{aligned}
&\left| \int_0^t \int_{\mathbb{R}^3} \left[P_{M,N}X_s^{M,N,(2)} - \frac{1}{\nu} P_{M,N}(\nu X_s^{M,N,(2)}) \right] (P_{M,N}X_s^{M,N,(2)})^3 \nu dx ds \right| \\
&\le CM^\sigma 2^{-N} \int_0^t \left\|X_s^{M,N,(2)}\right\|_{L^4(\nu)} \left\|P_{M,N}X_s^{M,N,(2)}\right\|_{L^4(\nu)}^3 ds.
\end{aligned}$$

Hence, by Hölder's inequality we have

$$\begin{aligned}
&- \int_0^t \int_{\mathbb{R}^3} X_s^{M,N,(2)} P_{M,N}^* \left[\left(P_{M,N}X_s^{M,N,(2)}\right)^3 \right] \nu dx ds \\
&\le -\frac{1}{2} \int_0^t \left\|P_{M,N}X_s^{M,N,(2)}\right\|_{L^4(\nu)}^4 ds + CM^{4\sigma}2^{-4N} \int_0^t \left\|X_s^{M,N,(2)}\right\|_{L^4(\nu)}^4 ds.
\end{aligned} \tag{6.9}$$

Since the integration by parts formula and Hölder's inequality imply

$$\begin{aligned}
&\int_0^t \int_{\mathbb{R}^3} X_s^{M,N,(2),\geqslant} \left[\left(\triangle - m_0^2\right) X_s^{M,N,(2),\geqslant} \right] \nu dx ds \\
&\le - \int_0^t \int_{\mathbb{R}^3} |\nabla X_s^{M,N,(2),\geqslant}|^2 \nu dx ds - m_0^2 \int_0^t \int_{\mathbb{R}^3} \left(X_s^{M,N,(2),\geqslant}\right)^2 \nu dx ds \\
&\quad - \int_0^t \int_{\mathbb{R}^3} X_s^{M,N,(2),\geqslant} \nabla X_s^{M,N,(2),\geqslant} \cdot (\nabla \log \nu) \nu dx ds \\
&\le -\left(1 - \frac{\sigma^2}{2m_0^2}\right) \int_0^t \int_{\mathbb{R}^3} |\nabla X_s^{M,N,(2),\geqslant}|^2 \nu dx ds - \frac{m_0^2}{2} \int_0^t \int_{\mathbb{R}^3} \left(X_s^{M,N,(2),\geqslant}\right)^2 \nu dx ds,
\end{aligned}$$

from (4.1) we have

$$
\begin{aligned}
&\int_0^t \int_{\mathbb{R}^3} X_s^{M,N,(2),\geqslant} \left[\left(\triangle - m_0^2\right) X_s^{M,N,(2),\geqslant} \right] \nu dx ds \\
&\leq -c \int_0^t \int_{\mathbb{R}^3} \left[|\nabla X_s^{M,N,(2),\geqslant}|^2 + \left(X_s^{M,N,(2),\geqslant} \right)^2 \right] \nu dx ds
\end{aligned} \tag{6.10}
$$

where c is a positive constant independent of M and N.

In view of Proposition 7.2, (6.9) and (6.10), by following the proof of [**AK20**, Proposition 4.11] with respect to $L^2(\nu)$ instead of the L^2-space on the torus, we obtain the assertion. □

PROPOSITION 6.6. For $\alpha \in [0, 1/2)$, $\gamma \in (0, 1/8)$, $\eta \in [0, 1)$, $\varepsilon \in (0, \gamma/4)$ and $t \in [0, T]$,

$$
\begin{aligned}
&\sup_{s',t' \in [0,t]; s' < t'} \frac{(s')^\eta \left\| X_{t'}^{M,N,(2)} - X_{s'}^{M,N,(2)} \right\|_{B_{4/3}^\alpha(\nu)}}{(t' - s')^\gamma} \\
&\leq K_{M,N} \sup_{r \in [0,t]} \left(r^\eta \left\| X_r^{M,N,(2),\geqslant} \right\|_{B_{4/3}^{\alpha+2\gamma}(\nu)} \right) \\
&\quad + K_{M,N} \sup_{r \in [0,t]} \left(r^\eta \left\| X_r^{M,N,(2),<} \right\|_{B_{4/3}^{\alpha+2\gamma}(\nu)} \right) \\
&\quad + K_{M,N} \int_0^t \left(\left\| P_{M,N} X_s^{M,N,(2)} \right\|_{L^4(\nu)}^4 + \left\| P_{M,N} X_s^{M,N,(2)} \right\|_{B_2^{15/16}(\nu)}^2 \right)^{7/8} ds \\
&\quad + K_{M,N} \mathfrak{Y}_\varepsilon^N(t) + K_{M,N} Q.
\end{aligned}
$$

PROOF. The proof is obtained in the same way as the proof of [**Kus20**, Proposition 3.1], which is an improved version of [**AK20**, Proposition 4.13]. See also Remark 6.3. □

PROPOSITION 6.7. For $\gamma \in (0, 1/8)$, $\eta \in [0, 1)$, $\varepsilon \in (0, \gamma/4)$, $q \in (1, 8/7)$, $t \in [0, T]$ and $\delta \in (0, 1]$,

$$
\mathfrak{Y}_\varepsilon^{M,N}(t) \leq C \left\| X_0^{M,N,(2)} \right\|_{B_{4/3}^{-1+2\gamma+3\varepsilon}(\nu)} + \delta K_{M,N} \mathfrak{X}_{\lambda,\alpha,\eta,\gamma}^{M,N}(t)^{7/8} + C_\delta K_{M,N} Q.
$$

PROOF. The proof is almost the same as that of [**AK20**, Proposition 4.15]. Indeed, by following the way that we obtained (4.17) in the proof of [**AK20**, Proposition 4.15], and by taking δ sufficiently small we obtain the estimate. □

PROPOSITION 6.8. For $\gamma \in (0, 1/8)$, $\alpha \in [0, (1 - 4\gamma)/2)$, $\eta \in [0, 1)$, $\varepsilon \in (0, \gamma/4)$, $q \in (1, 8/7)$, $t \in [0, T]$ and $\delta \in (0, 1]$, we have

$$
\begin{aligned}
&\sup_{r \in [0,t]} \left(r^\eta \left\| X_r^{M,N,(2),<} \right\|_{B_4^{\alpha+2\gamma}(\nu)}^3 \right) + \sup_{r \in [0,t]} \left(r^\eta \left\| X_r^{M,N,(2),\geqslant} \right\|_{B_{4/3}^{\alpha+2\gamma}(\nu)} \right) \\
&\leq C \left\| X_0^{M,N,(2)} \right\|_{B_{4/3}^{\alpha+2(\gamma-\eta)}(\nu)} + \delta K_{M,N} \mathfrak{X}_{\lambda,\alpha,\eta,\gamma}^{M,N}(t) + \delta K_{M,N} \mathfrak{Y}_\varepsilon^{M,N}(t)^q + C_\delta K_{M,N} Q.
\end{aligned}
$$

PROOF. The assertion is proved by following the proof of [**Kus20**, Proposition 3.2] without taking the expectation, which is an improved version of [**AK20**, Proposition 4.17]. □

PROPOSITION 6.9. For $\alpha \in [0,1/2)$, $\gamma \in (0,1/8)$, $\eta \in [0,1)$, $\varepsilon \in (0,\gamma/4)$ and $q \in (1,8/7)$, we have

$$\begin{aligned}
&\left\|X_T^{M,N,(2)}\right\|_{L^2(\nu)}^2 + \mathfrak{X}_{\lambda,\alpha,\eta,\gamma}^{M,N}(T) + \mathfrak{Y}_\varepsilon^{M,N}(T)^q \\
&+ \sup_{r\in[0,T]}\left(r^\eta \left\|X_r^{M,N,(2),<}\right\|_{B_4^{\alpha+2\gamma}(\nu)}^3\right) + \sup_{r\in[0,T]}\left(r^\eta \left\|X_r^{M,N,(2),\geqslant}\right\|_{B_{4/3}^{\alpha+2\gamma}(\nu)}\right) \\
&\le \left\|X_0^{M,N,(2)}\right\|_{L^2(\nu)}^2 + C\left\|X_0^{M,N,(2)}\right\|_{B_{4/3}^{-1+2\gamma+3\varepsilon}(\nu)}^q + C\left\|X_0^{M,N,(2)}\right\|_{B_{4/3}^{\alpha+2(\gamma-\eta)}(\nu)} \\
&\quad + CM^{4\sigma}2^{-4N}\int_0^T \left\|X_s^{M,N,(2)}\right\|_{L^4(\nu)}^4 ds + K_{M,N}Q.
\end{aligned}$$

PROOF. Propositions 6.5, 6.6, 6.7 and 6.8, and Hölder's inequality imply that for $\delta \in (0,1]$

$$\begin{aligned}
&\|X_T^{M,N,(2)}\|_{L^2(\nu)}^2 + \mathfrak{X}_{\lambda,\alpha,\eta,\gamma}^{M,N}(T) + \mathfrak{Y}_\varepsilon^{M,N}(T)^q \\
&+ \sup_{r\in[0,T]}\left(r^\eta \left\|X_r^{M,N,(2),<}\right\|_{B_4^{\alpha+2\gamma}(\nu)}^3\right) + \sup_{r\in[0,T]}\left(r^\eta \left\|X_r^{M,N,(2),\geqslant}\right\|_{B_{4/3}^{\alpha+2\gamma}(\nu)}\right) \\
&\le \delta \sup_{r\in[0,T]}\left(r^\eta \left\|X_r^{M,N,(2),\geqslant}\right\|_{B_{4/3}^{\alpha+2\gamma}(\nu)}\right) + \delta \sup_{r\in[0,T]}\left(r^\eta \left\|X_r^{M,N,(2),<}\right\|_{B_{4/3}^{\alpha+2\gamma}(\nu)}\right) \\
&\quad + \delta\|X_T^{M,N,(2),<}\|_{L^2(\nu)}^3 + \delta K_{M,N}\mathfrak{X}_{\lambda,\alpha,\eta,\gamma}^{M,N}(T) + \delta K_{M,N}\mathfrak{Y}_\varepsilon^{M,N}(T)^q \\
&\quad + \left\|X_0^{M,N,(2)}\right\|_{L^2(\nu)}^2 + C\left\|X_0^{M,N,(2)}\right\|_{B_{4/3}^{-1+2\gamma+3\varepsilon}(\nu)}^q + C\left\|X_0^{M,N,(2)}\right\|_{B_{4/3}^{\alpha+2(\gamma-\eta)}(\nu)} \\
&\quad + CM^{4\sigma}2^{-4N}\int_0^T \left\|X_s^{M,N,(2)}\right\|_{L^4(\nu)}^4 ds + C_\delta K_{M,N}Q.
\end{aligned}$$

Hence, by applying the fact that

$$\|X_T^{M,N,(2),<}\|_{L^2(\nu)}^3 \le T^{-\eta}\sup_{r\in[0,T]}\left(r^\eta \left\|X_r^{M,N,(2),<}\right\|_{B_4^{2\gamma}(\nu)}^3\right),$$

and taking δ sufficiently small we have the assertion. □

Now we prove the uniform estimate, which is applied to prove the tightness.

THEOREM 6.10. *Let* $\alpha \in [0,1/2)$, $\gamma \in (0,1/8)$, $\eta \in [0,1)$, $\varepsilon \in (0,\gamma/4)$ *and* $q \in (1,8/7)$. *Assume that* (4.1) *and*

$$\alpha + 2\gamma < \min\left\{2\eta + \frac{1}{2}, \frac{1-\varepsilon}{2}\right\}.$$

Then,

$$
\begin{aligned}
&E\left[\mathfrak{X}^{M,N}_{\lambda,\alpha,\eta,\gamma}(T)^{1/3}\right] + E\left[\mathfrak{Y}^{M,N}_{\varepsilon}(T)^{q/3}\right] \\
&+ E\left[\sup_{r\in[0,T]}\left(r^\eta \left\|X_r^{M,N,(2),<}\right\|^3_{B_4^{\alpha+2\gamma}}\right)^{1/3}\right] \\
&+ E\left[\sup_{r\in[0,T]}\left(r^\eta \left\|X_r^{M,N,(2),\geqslant}\right\|_{B_{4/3}^{\alpha+2\gamma}}\right)^{1/3}\right] \\
&\leq K_{M,N}.
\end{aligned}
$$

PROOF. It is sufficient to show the estimate for some large $T \in (0,\infty)$. Proposition 6.9, subadditivity of $x \mapsto x^{1/3}$ for $x \geq 0$ and the nonnegativity of each term imply

$$
\begin{aligned}
&\mathfrak{X}^{M,N}_{\lambda,\alpha,\eta,\gamma}(T)^{1/3} + \mathfrak{Y}^{M,N}_{\varepsilon}(T)^{q/3} \\
&+ \sup_{r\in[0,T]}\left(r^\eta \left\|X_r^{M,N,(2),<}\right\|^3_{B_4^{\alpha+2\gamma}(\nu)}\right)^{1/3} + \sup_{r\in[0,T]}\left(r^\eta \left\|X_r^{M,N,(2),\geqslant}\right\|_{B_{4/3}^{\alpha+2\gamma}(\nu)}\right)^{1/3} \\
&\leq 4\left\|X_0^{M,N,(2)}\right\|^{2/3}_{L^2(\nu)} + C\left\|X_0^{M,N,(2)}\right\|^{q/3}_{B_{4/3}^{-1+2\gamma+3\varepsilon}(\nu)} + C\left\|X_0^{M,N,(2)}\right\|^{1/3}_{B_{4/3}^{\alpha+2(\gamma-\eta)}(\nu)} \\
&\quad + CM^{4\sigma/3}\left(2^{-4N}\int_0^T \left\|X_s^{M,N,(2)}\right\|^4_{L^4(\nu)} ds\right)^{1/3} + K_{M,N}Q.
\end{aligned}
$$

But the Besov embedding theorem, Corollaries 5.3 and 5.5 yield

$$
\begin{aligned}
&M^{4\sigma/3}E\left[\left(2^{-4N}\int_0^T \left\|X_s^{M,N,(2)}\right\|^4_{L^4(\nu)} ds\right)^{1/3}\right] \\
&\leq CM^{4\sigma/3}2^{-(4/3)N}E\left[\left(\int_0^T \left\|X_s^{M,N,(2)}\right\|^4_{B_2^{3/4}} ds\right)^{1/3}\right] \\
&\leq CM^{4\sigma/3}2^{-(4/3)N}E\left[\left(\int_0^T \left\|X_s^{M,N,(2)}\right\|^2_{B_2^{1/2}}\left\|X_s^{M,N,(2)}\right\|^2_{B_2^{1}} ds\right)^{1/3}\right] \\
&\leq CM^{4\sigma/3}2^{-(4/3)N}E\left[\left(\sup_{s\in[0,T]}\left\|X_s^{M,N,(2)}\right\|^{4/3}_{B_2^{1/2}}\right)^{1/2}\left(\int_0^t \left\|X_s^{M,N,(2)}\right\|^2_{B_2^{1}} ds\right)^{1/3}\right] \\
&\leq CM^{4\sigma/3}2^{-(4/3)N}E\left[\sup_{s\in[0,T]}\left\|X_s^{M,N,(2)}\right\|^{4/3}_{B_2^{1/2}}\right]^{1/2} \\
&\qquad\qquad \times\left(1 + E\left[\int_0^t \left\|X_s^{M,N,(2)}\right\|^2_{B_2^{1}} ds\right]^{2/3}\right) \\
&\leq K_{M,N},
\end{aligned}
$$

so that we obtain

$$
\begin{aligned}
&E\left[\mathfrak{X}^{M,N}_{\lambda,\alpha,\eta,\gamma}(T)^{1/3}\right] + E\left[\mathfrak{Y}^{M,N}_{\varepsilon}(T)^{q/3}\right] \\
&+E\left[\sup_{r\in[0,T]}\left(r^{\eta}\left\|X^{M,N,(2),<}_{r}\right\|^{3}_{B^{\alpha+2\gamma}_{4}(\nu)}\right)^{1/3}\right] \\
&+E\left[\sup_{r\in[0,T]}\left(r^{\eta}\left\|X^{M,N,(2),\geqslant}_{r}\right\|_{B^{\alpha+2\gamma}_{4/3}(\nu)}\right)^{1/3}\right] \\
&\leq 4E\left[\left\|X^{M,N,(2)}_{0}\right\|^{2/3}_{L^{2}(\nu)}\right] + CE\left[\left\|X^{M,N,(2)}_{0}\right\|^{q/3}_{B^{-1+2\gamma+3\varepsilon}_{4/3}(\nu)}\right] \\
&\quad +CE\left[\left\|X^{M,N,(2)}_{0}\right\|^{1/3}_{B^{\alpha+2(\gamma-\eta)}_{4/3}(\nu)}\right] + K_{M,N}.
\end{aligned} \tag{6.11}
$$

From the stationarity of the marginal laws of $X^{M,N,(1)}$ we have, for any $T > 0$:

$$
\begin{aligned}
&E\left[\left\|X^{M,N,(2)}_{0}\right\|^{2/3}_{L^{2}(\nu)}\right] \\
&\leq \frac{1}{T}\int_{0}^{T} E\left[\left\|X^{M,N,(1)}_{t}\right\|^{2/3}_{L^{2}(\nu)}\right]dt + \lambda^{2/3}E\left[\left\|\mathcal{Z}^{(0,3,M,N)}_{0}\right\|^{2/3}_{L^{2}(\nu)}\right] \\
&\leq T^{-1/3}E\left[\left(\int_{0}^{T}\left\|X^{M,N,(2)}_{t}\right\|^{2}_{L^{2}(\nu)}dt\right)^{1/3}\right] + 2\lambda^{2/3}\sup_{t\in[0,T]}E\left[\left\|\mathcal{Z}^{(0,3,M,N)}_{t}\right\|^{2/3}_{L^{2}(\nu)}\right] \\
&\leq T^{-1/3}E\left[\mathfrak{X}^{M,N}_{\lambda,\alpha,\eta,\gamma}(T)^{1/3}\right] + C.
\end{aligned}
$$

Hence, by taking T sufficiently large and applying (6.11) we have

$$
\begin{aligned}
&E\left[\mathfrak{X}^{M,N}_{\lambda,\alpha,\eta,\gamma}(T)^{1/3}\right] + E\left[\mathfrak{Y}^{M,N}_{\varepsilon}(T)^{q/3}\right] \\
&+E\left[\sup_{r\in[0,T]}\left(r^{\eta}\left\|X^{M,N,(2),<}_{r}\right\|^{3}_{B^{\alpha+2\gamma}_{4}(\nu)}\right)^{1/3}\right] \\
&+E\left[\sup_{r\in[0,T]}\left(r^{\eta}\left\|X^{M,N,(2),\geqslant}_{r}\right\|_{B^{\alpha+2\gamma}_{4/3}(\nu)}\right)^{1/3}\right] \\
&\leq CE\left[\left\|X^{M,N,(2)}_{0}\right\|^{q/3}_{B^{-1+2\gamma+3\varepsilon}_{4/3}(\nu)}\right] + CE\left[\left\|X^{M,N,(2)}_{0}\right\|^{1/3}_{B^{\alpha+2(\gamma-\eta)}_{4/3}(\nu)}\right] \\
&\quad +K_{M,N}.
\end{aligned} \tag{6.12}
$$

The stationarity of the marginal laws of $X^{M,N,(1)}$ also implies that

$$\begin{aligned}
&E\left[\left\|X_0^{M,N,(2)}\right\|_{B_{4/3}^{-1+2\gamma+3\varepsilon}(\nu)}^{q/3}\right] \\
&\leq \frac{1}{T}\int_0^T E\left[\left\|X_t^{M,N,(1)}\right\|_{B_{4/3}^{-1+2\gamma+3\varepsilon}(\nu)}^{q/3}\right] dt + \lambda^{q/3} E\left[\left\|\mathcal{Z}_0^{(0,3,M,N)}\right\|_{B_{4/3}^{-1+2\gamma+3\varepsilon}(\nu)}^{q/3}\right] \\
&\leq CE\left[\left(\int_0^T \left\|X_t^{M,N,(2)}\right\|_{B_{4/3}^{-1+2\gamma+3\varepsilon}(\nu)}^{q} dt\right)^{1/3}\right] \\
&\qquad + 2\lambda^{q/3} \sup_{t\in[0,T]} E\left[\left\|\mathcal{Z}_t^{(0,3,M,N)}\right\|_{B_{4/3}^{-1+2\gamma+3\varepsilon}(\nu)}^{q/3}\right] \\
&\leq CE\left[\left(\delta^3 \mathfrak{X}_{\lambda,\alpha,\eta,\gamma}^{M,N}(T) + C_\delta\right)^{1/3}\right] + C \\
&\leq C\delta E\left[\mathfrak{X}_{\lambda,\alpha,\eta,\gamma}^{M,N}(T)^{1/3}\right] + C_\delta.
\end{aligned}$$

Similarly it holds that

$$E\left[\left\|X_0^{M,N,(2)}\right\|_{B_{4/3}^{\alpha+2(\gamma-\eta)}(\nu)}^{1/3}\right] \leq C\delta E\left[\mathfrak{X}_{\lambda,\alpha,\eta,\gamma}^{M,N}(T)^{1/3}\right] + C_\delta.$$

By taking a sufficiently small δ in these inequalities and combining them with (6.12) we obtain the assertion. □

REMARK 6.11. The assumption on η is improved from those in [**AK20**] and [**Kus20**]. This improvement comes from applying the stationarity of $X^{M,N,(1)} = X^{M,N} - Z$ instead of that of $X^{M,N}$ (Z is the Ornstein-Uhlenbeck process of Chapter 3).

Theorem 6.10 yields the tightness of the laws of $\{X^{M,N}\}$, which was the target in the present paper. Now, by using Theorem 6.10 we manage to prove our key theorem (Theorem 1.1).

PROOF OF THEOREM 1.1. Choose σ, α, γ, η, ε and q so that they satisfy the assumptions in Theorem 6.10. Then, in view of Proposition 6.9 and Theorem 6.10, we have the bound

$$\begin{aligned}
&E\left[\mathfrak{X}_{\lambda,\alpha,\eta,\gamma}^{M_N,N}(T)^{1/3}\right] + E\left[\mathfrak{Y}_\varepsilon^{M_N,N}(T)^{q/3}\right] \\
&+E\left[\sup_{r\in[0,T]}\left(r^\eta \left\|X_r^{M_N,N,(2),<}\right\|_{B_4^{\alpha+2\gamma}(\nu)}^3\right)^{1/3}\right] \\
&+E\left[\sup_{r\in[0,T]}\left(r^\eta \left\|X_r^{M_N,N,(2),\geqslant}\right\|_{B_{4/3}^{\alpha+2\gamma}(\nu)}\right)^{1/3}\right] \leq C
\end{aligned} \tag{6.13}$$

for some positive constant C. Let $T \in (0,\infty)$ and $t_0 \in (0,T)$. Then, (6.13) and Chebyshev's inequality imply that, for any $\varepsilon' > 0$

$$\begin{aligned}
&\limsup_{h\downarrow 0} \sup_{N\in\mathbb{N}} P\left(\sup_{s,t\in[t_0,T];|s-t|<h} \left\| X_t^{M_N,N,(2)} - X_s^{M_N,N,(2)} \right\|_{B^\alpha_{4/3}(\nu)} > \varepsilon' \right) \\
&\le \limsup_{h\downarrow 0} \left(\frac{h^\gamma}{\varepsilon' t_0^\eta} \right)^{1/3} \\
&\qquad \times \sup_{N\in\mathbb{N}} E\left[\left(\sup_{\substack{s,t\in[t_0,T];\\ s<t,t-s<h}} \frac{s^\eta \left\| X_t^{M_N,N,(2)} - X_s^{M_N,N,(2)} \right\|_{B^\alpha_{4/3}(\nu)}}{(t-s)^\gamma} \right)^{1/3} \right] \\
&= 0,
\end{aligned}$$

moreover:

$$\begin{aligned}
&\limsup_{R\to\infty} \sup_{N\in\mathbb{N}} P\left(\left\| X_{t_0}^{M_N,N,(2)} \right\|_{B^{\alpha+2\gamma}_{4/3}(\nu)} > R \right) \\
&\le \limsup_{R\to\infty} \frac{1}{R^{1/3} t_0^\eta} \sup_{N\in\mathbb{N}} E\left[\sup_{r\in[0,T]} \left(r^\eta \left\| X_r^{M_N,N,(2)} \right\|_{B^{\alpha+2\gamma}_{4/3}(\nu)} \right)^{1/3} \right] = 0.
\end{aligned}$$

In view of Proposition 2.6, we have then tightness of the laws of $X^{M_N,N,(2)}$ as probability measures on $C([t_0,T]; B^\alpha_q(\nu))$ for $\alpha \in [0,1/2)$ and $q \in [1,4/3)$ (c.f. [**IW89**, Theorem 4.2 of Chapter I]). Hence, an argument similar to the one in the proof of [**AK20**, Theorem 4.19] yields the tightness of the laws of $\{X^{M_N,N}\}$ on $C([0,\infty); B_q^{-1/2-\varepsilon}(\nu))$ for $q \in [1,4/3)$. The embedding theorem of weighted Besov spaces (see Proposition 2.4) implies the tightness of the laws of $\{X^{M_N,N}\}$ on $C([0,\infty); B_{4/3}^{-1/2-\varepsilon-\delta}(\nu^{4/(3q)}))$ where $\delta = 3[(1/q) - (3/4)]$. Noting that

$$\nu^{4/(3q)}(x) = (1+|x|^2)^{-2\sigma/(3q)}$$

and $\varepsilon \in (0,1)$, $q \in [1,4/3)$ and $\sigma \in (3, \sqrt{2m_0^2})$ are arbitrary (see (4.1)), we obtain the tightness of the laws of $\{X^{M_N,N}\}$ on $C([0,\infty); B_{4/3}^{-1/2-\varepsilon}(\nu))$ for all $\varepsilon \in (0,1/16]$ and $\nu(x) = (1+|x|^2)^{-\sigma/2}$ with σ satisfying (4.1).

Again by the tightness of the laws of $X^{M_N,N,(2)}$ on $C([t_0,T]; B^\alpha_q(\nu))$ for $\alpha \in [0,1/2)$ and $q \in [1,4/3)$ and Proposition 2.4 we get the tightness of the laws of $X^{M_N,N,(2)}$ on $C([t_0,T]; B_p^{-1/2-\varepsilon}(\nu^{p/q}))$, where $p \in [1,\infty)$ is restricted by

$$\alpha = -\frac{1}{2} - \varepsilon + 3\left(\frac{1}{q} - \frac{1}{p} \right).$$

Since $\varepsilon \in (0,1/16]$, $\alpha \in [0,1/2)$ and $q \in [1,4/3)$ are arbitrary, we have the tightness of the laws of $X^{M_N,N,(2)}$ on $C([t_0,T]; B_{12/5}^{-1/2-\varepsilon}(\nu^{9/5}))$ for all $\varepsilon \in (0,1/16]$ and $\nu(x) = (1+|x|^2)^{-\sigma/2}$ with σ satisfying (4.1). Again similarly to the proof of [**AK20**, Theorem 4.19] (see also the proof of [**Kus20**, Theorem 3.4]), we have the tightness of the laws of $X^{M_N,N}$ on $C([0,\infty); B_{12/5}^{-1/2-\varepsilon}(\nu^{9/5}))$ with ε and ν as in Theorem 1.1. Thus, we obtain the tightness of the laws of $X^{M_N,N}$ on $C([0,\infty); B_{4/3}^{-1/2-\varepsilon}(\nu) \cap$

$B^{-1/2-\varepsilon}_{12/5}(\nu^{9/5})$). Moreover, the tightness holds for all $\sigma \in (3, \infty)$, because the weighted Besov spaces have the monotonicity property in σ.

The proofs of the other assertions run in the same way as in the proof of [**AK20**, Theorem 4.19]. So, we omit them. □

CHAPTER 7

Properties of the constructed Φ^4_3-measure

7.1. Rotation and reflection invariance

In this section, we prove Theorem 1.2(i). Define ψ_N, ρ_M, $P_{M,N}$, $\mu_{M,N}$ and $Z_{M,N}$ as in Section 1.2. Here, we remark that $Z_{M,N}$ is the normalizing constant. The proofs are very simple, because we are able to prove $\mu_{M,N}$ is rotation invariant if ψ and ρ are radial functions as follows.

PROOF OF THEOREM 1.2(i). Assume that ψ and ρ are radial (i.e. rotation invariant) functions. Then, so are ψ_N and ρ_M. Hence, we have $U_{M,N}(\phi) = U_{M,N}(\theta\phi)$ for any rotation θ acting in $\mathbb{R}^3$. This fact and the rotation and reflection invariance of the free field measure μ_0 yield the rotation and reflection invariance of $\mu_{M,N}$. Therefore, by taking the limit of the subsequence $\mu_{M_{N(k)},N(k)}$ for $k \to \infty$ with $M_{N(k)}, N(k) \to \infty$, which converges to μ, we obtain Theorem 1.2(i). $\square$

7.2. Reflection positivity of the limit measure

In this section we prove Theorem 1.2(ii). In the proof we show that the image measure $P_N\mu_{M,N}$ of $\mu_{M,N}$ with respect to P_N has the reflection positivity property by means of the Markov property of $P_N\phi$ under the free field μ_0 for each N.

PROOF OF THEOREM 1.2(ii). Let $\{g_\varepsilon\}_{\varepsilon>0}$ be a sequence in $C_b^\infty(\mathbb{R}^3)$ such that the support of g_ε is included in $\{(x_1,x_2,x_3); x_1 > \varepsilon\}$, $\sup_{\varepsilon>0}\|g_\varepsilon\|_{L^\infty} \le 1$ and $\lim_{\varepsilon\downarrow 0} g_\varepsilon(x) = \mathbb{I}_{(0,\infty)}(x_1)$ for $x=(x_1,x_2,x_3)\in\mathbb{R}^3$, and let

$$
\begin{aligned}
&U^{\varepsilon,+}_{M,N}(\phi)\\
&:= \int_{\mathbb{R}^3}\left[\frac{\lambda}{4}(g_\varepsilon P_{M,N}\phi)^4 - \frac{3\lambda}{2}\left(C_1^{(N)} - 3\lambda C_2^{(M,N)}\right)\rho_M^2 g_\varepsilon^2 (g_\varepsilon P_{M,N}\phi)^2\right]dx,\\
&U^{\varepsilon,-}_{M,N}(\phi)\\
&:= \int_{\mathbb{R}^3}\left[\frac{\lambda}{4}(g_\varepsilon^\dagger P_{M,N}\phi)^4 - \frac{3\lambda}{2}\left(C_1^{(N)} - 3\lambda C_2^{(M,N)}\right)\rho_M^2 (g_\varepsilon^\dagger)^2 (g_\varepsilon^\dagger P_{M,N}\phi)^2\right]dx.
\end{aligned}
$$

Denote by $\mu_0(d\phi|(P_N\phi)_{x_1=0})$ the regular conditional probability measure of μ_0 with respect to the σ-field generated by

$$
\left\{\langle f, P_N\phi\rangle; f\in\mathscr{H}^{-1}(\mathbb{R}^3),\ \operatorname{supp} f\subset\{(x_1,x_2,x_3); x_1=0\}\right\}.
$$

Here, note that $\mathscr{H}^{-1}(\mathbb{R}^3)$ is the Sobolev space generated by the norm $\|f\|_{\mathscr{H}^{-1}(\mathbb{R}^3)} = \|(m_0^2-\triangle)^{-1/2}f\|_{L^2(\mathbb{R}^3,dx)}$, which coincides with the Cameron-Martin space of the free field measure μ_0. Similarly to the proof of [**Nel73b**, Theorem 5], we can prove that $P_N\phi$ under the free field measure μ_0 is a Markov field.[1] Indeed, even if we replace $\mathscr{H}^{-1}(\mathbb{R}^3)$ by a closed linear subspace

$$\mathscr{H}_N^{-1}(\mathbb{R}^3) := \left\{P_N f;\ f \in \mathscr{H}^{-1}(\mathbb{R}^3)\right\},$$

the proof goes in a similar way. Moreover, $g_\varepsilon P_{M,N}\phi$ and $g_\varepsilon^\dagger P_{M,N}\phi$ are measurable with respect to the σ-field generated by $\{\langle f, P_{M,N}\phi\rangle; \operatorname{supp} f \subset \{(x_1,x_2,x_3); x_1 > 0\}\}$ and $\{\langle f, P_{M,N}\phi\rangle; \operatorname{supp} f \subset \{(x_1,x_2,x_3); x_1 < 0\}\}$, respectively. Hence, by these facts and the assumption on F, we have

$$\begin{aligned}
&\int \overline{F^\dagger(P_N\phi)} F(P_N\phi) \exp(-U^{\varepsilon,+}_{M,N}(\phi) - U^{\varepsilon,-}_{M,N}(\phi))\mu_0(d\phi) \\
&= \int \left(\int \overline{F^\dagger(P_N\phi)} \exp(-U^{\varepsilon,-}_{M,N}(\phi))\mu_0(d\phi|(P_N\phi)_{x_1=0}) \right) \\
&\qquad \times \left(\int F(P_N\phi) \exp(-U^{\varepsilon,+}_{M,N}(\phi))\mu_0(d\phi|(P_N\phi)_{x_1=0}) \right) \mu_0(d\phi).
\end{aligned} \tag{7.1}$$

Note that $C_2^{(M,N)\dagger} = C_2^{(M,N)}$, because $\delta_{-x}(\cdot) = \delta_x(-\cdot)$, $\rho^\dagger = \rho$ and $\psi^\dagger = \psi$ imply

$$\begin{aligned}
&\left\langle (m_0^2 - \triangle)^{-1} e^{t(\triangle - m_0^2)} \left(\rho_M^2 \Delta_i \delta_{-x}\right), \rho_M^2 e^{t(\triangle - m_0^2)} P_N^2 \Delta_j \delta_{-x} \right\rangle \\
&= \left\langle (m_0^2 - \triangle)^{-1} e^{t(\triangle - m_0^2)} \left(\rho_M^2 \Delta_i \delta_{x}\right), \rho_M^2 e^{t(\triangle - m_0^2)} P_N^2 \Delta_j \delta_{x} \right\rangle.
\end{aligned}$$

We also note that $U^{\varepsilon,-}_{M,N}(\phi^\dagger) = U^{\varepsilon,+}_{M,N}(\phi)$ and $(P_N\phi)^\dagger = P_N(\phi^\dagger)$ follow in a similarly way. Since the laws of $(P_N\phi)^\dagger$ and $P_N\phi$ under $\mu_0(d\phi|(P_N\phi)_{x_1=0})$ are same,

$$\begin{aligned}
&\int \overline{F^\dagger(P_N\phi)} \exp(-U^{\varepsilon,-}_{M,N}(\phi))\mu_0(d\phi|(P_N\phi)_{x_1=0}) \\
&= \int \overline{F^\dagger(P_N\phi^\dagger)} \exp(-U^{\varepsilon,-}_{M,N}(\phi^\dagger))\mu_0(d\phi|(P_N\phi)_{x_1=0}) \\
&= \int \overline{F(P_N\phi)} \exp(-U^{\varepsilon,+}_{M,N}(\phi))\mu_0(d\phi|(P_N\phi)_{x_1=0}).
\end{aligned}$$

Thus, we have

$$\begin{aligned}
&\int \overline{F^\dagger(P_N\phi)} F(P_N\phi) \exp(-U^{\varepsilon,+}_{M,N}(\phi) - U^{\varepsilon,-}_{M,N}(\phi))\mu_0(d\phi) \\
&= \int \left| \int F(P_N\phi) \exp(-U^{\varepsilon,+}_{M,N}(\phi))\mu_0(d\phi|(P_N\phi)_{x_1=0}) \right|^2 \mu_0(d\phi) \geq 0.
\end{aligned}$$

This inequality and the fact that

$$\begin{aligned}
&\lim_{\varepsilon\downarrow 0} \left(U^{\varepsilon,+}_{M,N}(\phi) + U^{\varepsilon,-}_{M,N}(\phi) \right) \\
&= \int_{\mathbb{R}^3} \left[\frac{\lambda}{4}(P_{M,N}\phi)^4 - \frac{3\lambda}{2}\left(C_1^{(N)} - 3\lambda C_2^{(M,N)}\right) \rho_M^2 (P_{M,N}\phi)^2 \right] dx
\end{aligned}$$

imply that

$$\int \overline{F^\dagger(P_N\phi)} F(P_N\phi)\mu_{M,N}(d\phi) \geq 0.$$

[1] Markov field property of P_N under μ_0 based on the proof of [Nel73b, Theorem 5] is erroneous, because $\mathscr{H}_N^{-1}(\mathbb{R}^3)$ below is not a closed subspace in $\mathscr{H}^{-1}(\mathbb{R}^3)$. However, once we obtain (7.1), the rest of the proof works. A detailed modification will be left to future work.

Since F is bounded and Lipschitz continuous on B^s_p for $s \in \mathbb{R}$, $p \in [1, \infty)$, and $\mu_{M(k),N(k)}$ converges weakly to μ as $k \to \infty$,

$$\lim_{k\to\infty} \int \overline{F^\dagger(\phi)} F(\phi) \mu_{M_{N(k)},N(k)}(d\phi) = \int \overline{F^\dagger(\phi)} F(\phi) \mu(d\phi),$$
$$\left|F(P_{N(k)}\phi) - F(\phi)\right| + \left|\overline{F^\dagger(P_{N(k)}\phi)} - \overline{F^\dagger(\phi)}\right| \le C \min\{1, \|P_{N(k)}\phi - \phi\|_{B^s_p}\}$$

with a constant C. Therefore, we also obtain the reflection positivity of μ. $\square$

7.3. Support of the limit measure

In this section we prove Theorem 1.2(iii). Let ν, $X^{M,N}$, $X^{M,N,(1)}$, $X^{M,N,(2)}$, Z and $\mathcal{Z}^{(0,3,M,N)}$ be as in Chapter 4, and let X be the limit process of a subsequence $X^{M_{N(k)},N(k)}$ obtained in Theorem 1.1.

We choose α, γ, η, ε and q as in Theorem 6.10. By applying the Skorohod representation theorem to the pair of processes

$$(X^{M_{N(k)},N(k)}, \mathcal{Z}^{(1,M_{N(k)},N(k))}, \mathcal{Z}^{(0,3,M_{N(k)},N(k))}),$$

we have a sequence of the pair of processes on another probability space which does not change the laws and converges almost surely. For simplicity, we denote the almost-surely converging sequences and the limit by

$$\{X^{M_{N(k)},N(k)}\}, \{\mathcal{Z}^{(1,M_{N(k)},N(k))}\}, \{\mathcal{Z}^{(0,3,M_{N(k)},N(k))}\}$$

and X again. We remark that $X^{M_{N(k)},N(k),(2)}$ also converges almost surely on a suitable path space.

LEMMA 7.1. *For sufficiently small $\varepsilon \in (0,1)$, there exists a subsequence $\{N'(k)\}$ of $\{N(k)\}$ such that*

$$E\left[\limsup_{k\to\infty} \int_0^T \left\|X_s^{M_{N'(k)},N'(k),(2)}\right\|^2_{B_2^{1-\varepsilon}(\nu)} ds\right] < \infty,$$
$$E\left[\limsup_{k\to\infty} \int_0^T \left(\left\|X_s^{M_{N'(k)},N'(k)} - Z_s\right\|^2_{B_2^{1/2-\varepsilon}(\nu)} + \left\|X_s^{M_{N'(k)},N'(k)} - Z_s\right\|^4_{L^4(\nu)}\right) ds\right] < \infty.$$

PROOF. In this proof, for simplicity we denote the limits of $X^{M_{N(k)},N(k),(2)}$ and $\mathcal{Z}^{(0,3,M_{N(k)},N(k))}$ as $k \to \infty$ by $X^{(2)}$ and $\mathcal{Z}^{(0,3)}$, respectively. Then,

$$\lim_{k\to\infty} X_t^{M_{N(k)},N(k),(2)} = X_t^{(2)} - Z_t + \lambda \mathcal{Z}_t^{(0,3)}$$

in $B^{1/2-\varepsilon'}_{4/3}(\nu)$ uniformly in $t \in [0,T]$ for any $\varepsilon' \in (0,1]$. Proposition 6.9 implies

$$\begin{aligned}
&\int_0^T \left(\left\|X_s^{M_{N(k)},N(k),(2)}\right\|^2_{B_2^{1-\varepsilon}(\nu)} + \lambda \left\|P_{M_{N(k)},N(k)} X_s^{M_{N(k)},N(k),(2)}\right\|^4_{L^4(\nu)}\right) ds \\
&\le C \left\|X_0^{M_{N(k)},N(k),(2)}\right\|^2_{L^2(\nu)} + C \left\|X_0^{M_{N(k)},N(k),(2)}\right\|_{B_{4/3}^{\alpha+2(\gamma-\eta)}(\nu)} \\
&\quad + C2^{-(4-\varepsilon)N(k)} \int_0^T \left\|X_s^{M_{N(k)},N(k),(2)}\right\|^4_{L^4(\nu)} ds + CQ
\end{aligned} \tag{7.2}$$

for some positive constant C. Similarly to the proof of Theorem 6.10, we have the convergence

$$\lim_{N\to\infty} E\left[\left(2^{-(4-\varepsilon)N}\int_0^T \left\|X_s^{M_N,N,(2)}\right\|_{L^4(\nu)}^4 ds\right)^{1/3}\right] = 0.$$

In particular, by taking a subsequence $\{N'(k)\}$ of $\{N(k)\}$ we have

$$\lim_{k\to\infty} 2^{-(4-\varepsilon)N'(k)}\int_0^T \left\|X_s^{M_{N'(k)},N'(k),(2)}\right\|_{L^4(\nu)}^4 ds = 0 \tag{7.3}$$

almost surely. Hence (7.2), (7.3) and a generalized version of Fatou's lemma (see [**Sch71**]) imply

$$\begin{aligned} &E\left[\min\left\{\int_0^T \left(\left\|X_s^{(2)}\right\|_{B_2^{1-\varepsilon}(\nu)}^2 + \lambda\left\|X_s^{(2)}\right\|_{L^4(\nu)}^4\right) ds, n\right\}\right] \\ &\le CE\left[\min\left\{\left\|X_0^{(2)}\right\|_{L^2(\nu)}^2, n\right\}\right] + CE\left[\min\left\{\left\|X_0^{(2)}\right\|_{B_{4/3}^{\alpha+2(\gamma-\eta)}(\nu)}, n\right\}\right] + C. \end{aligned} \tag{7.4}$$

for any $n\in\mathbb{N}$. We now observe that $X_t - Z_t$ is a stationary process, because it is a limit of a stationary process (see the proof of [**AK20**, Theorem 4.19]). From this fact and Proposition 3.3 we have, for any $T>0$:

$$\begin{aligned} &E\left[\min\left\{\left\|X_0^{(2)}\right\|_{L^2(\nu)}^2, n\right\}\right] \\ &\le 2E\left[\min\left\{\|X_0 - Z_0\|_{L^2(\nu)}^2, n\right\}\right] + 2\lambda^2 E\left[\left\|\mathcal{Z}_0^{(0,3)}\right\|_{L^2(\nu)}^2\right] \\ &= \frac{2}{T}\int_0^T E\left[\min\left\{\|X_s - Z_s\|_{L^2(\nu)}^2, n\right\}\right] ds + C \\ &\le 2E\left[\min\left\{\frac{1}{T}\int_0^T \left\|X_s^{(2)}\right\|_{L^2(\nu)}^2 ds, n\right\}\right] + C \\ &\le 2E\left[\min\left\{\left(\frac{1}{T}\int_0^T \left\|X_s^{(2)}\right\|_{L^4(\nu)}^4 ds\right)^{1/2}, n\right\}\right] + C. \end{aligned}$$

Similarly, for any $n\in\mathbb{N}$ and some positive constant C:

$$\begin{aligned} &E\left[\min\left\{\left\|X_0^{(2)}\right\|_{B_{4/3}^{\alpha+2(\gamma-\eta)}(\nu)}, n\right\}\right] \\ &\le E\left[\min\left\{\left(\frac{1}{T}\int_0^T \left\|X_s^{(2)}\right\|_{B_2^{\alpha+2(\gamma-\eta)}(\nu)}^2 ds\right)^{1/2}, n\right\}\right] + C \\ &\le E\left[\min\left\{\left(\frac{1}{T}\int_0^T \left\|X_s^{(2)}\right\|_{B_2^{1-\varepsilon}(\nu)}^2 ds\right)^{1/2}, n\right\}\right] + C. \end{aligned}$$

These inequalities and (7.4) imply

$$E\left[\min\left\{\int_0^T\left(\left\|X_s^{(2)}\right\|_{B_2^{1-\varepsilon}(\nu)}^2+\lambda\left\|X_s^{(2)}\right\|_{L^4(\nu)}^4\right)ds,n\right\}\right]$$
$$\leq CE\left[\min\left\{\left(\int_0^T\left(\left\|X_s^{(2)}\right\|_{B_2^{1-\varepsilon}(\nu)}^2+\lambda\left\|X_s^{(2)}\right\|_{L^4(\nu)}^4\right)ds\right)^{1/2},n\right\}\right]+C.$$

Hence, by applying Proposition A.1 we have

$$E\left[\int_0^T\left(\left\|X_s^{(2)}\right\|_{B_2^{1-\varepsilon}(\nu)}^2+\lambda\left\|X_s^{(2)}\right\|_{L^4(\nu)}^4\right)ds\right]<\infty.$$

In particular, by the stationarity of X_t-Z_t and Proposition 3.3 we get

$$E\left[\left\|X_0^{(2)}\right\|_{L^2(\nu)}^4\right]+E\left[\left\|X_0^{(2)}\right\|_{B_{4/3}^{\alpha+2(\gamma-\eta)}(\nu)}^2\right]<\infty. \tag{7.5}$$

On the other hand, (7.2) and (7.3) also imply

$$E\left[\limsup_{k\to\infty}\int_0^T\left(\left\|X_s^{M_{N'(k)},N'(k),(2)}\right\|_{B_2^{1-\varepsilon}(\nu)}^2\right.\right.$$
$$\left.\left.+\lambda\left\|P_{M_{N'(k)},N'(k)}X_s^{M_{N'(k)},N'(k),(2)}\right\|_{L^4(\nu)}^4\right)ds\right]$$
$$\leq CE\left[\left\|X_0^{(2)}\right\|_{L^2(\nu)}^2\right]+CE\left[\left\|X_0^{(2)}\right\|_{B_{4/3}^{\alpha+2(\gamma-\eta)}(\nu)}\right]+C.$$

This inequality and (7.5) yield the first inequality. Applying Proposition 3.3 to (7.4), we obtain the second inequality. □

Proof of Theorem 1.2(iii). By the stationarity of $X^{M_N,N}$ and the fact that the marginal law X is equal to μ, for any $J\in\mathbb{N}$ we have

$$\int\sum_{j=-1}^{J}2^{-[(1/2)+\varepsilon]j}\|\Delta_j\phi\|_{L^p(\nu^{p/2})}^2\mu(d\phi)$$
$$=E\left[\int_0^1\sum_{j=-1}^{J}2^{-[(1/2)+\varepsilon]j}\|\Delta_jX_t\|_{L^p(\nu^{p/2})}^2dt\right]$$
$$=E\left[\liminf_{k\to\infty}\int_0^1\sum_{j=-1}^{J}2^{-[(1/2)+\varepsilon]j}\left\|\Delta_jX_t^{M_{N(k)},N(k)}\right\|_{L^p(\nu^{p/2})}^2dt\right]$$
$$\leq E\left[\liminf_{k\to\infty}\int_0^1\left\|\Delta_jX_t^{M_{N(k)},N(k)}\right\|_{B_p^{-1/2-\varepsilon}(\nu^{p/2})}^2dt\right]$$
$$\leq CE\left[\liminf_{k\to\infty}\int_0^1\left\|\Delta_jX_t^{M_{N(k)},N(k),(2)}\right\|_{B_p^{-1/2-\varepsilon}(\nu^{p/2})}^2dt\right]+C.$$

On the other hand, Proposition 2.4 implies

$$\left\|X_t^{M_{N(k)},N(k),(2)}\right\|_{B_p^{-1/2-\varepsilon}(\nu^{p/2})}\leq C\left\|X_t^{M_{N(k)},N(k),(2)}\right\|_{B_2^{1-\varepsilon}(\nu)}.$$

Hence, by the monotone convergence theorem and Fatou's lemma we have

$$\begin{aligned}
&\int \|\phi\|^2_{B_p^{-1/2-\varepsilon}(\nu^{p/2})}\mu(d\phi) \\
&\leq \liminf_{J\to\infty} \int \sum_{j=-1}^{J} 2^{-[(1/2)+\varepsilon]j} \|\Delta_j \phi\|^2_{L^p(\nu^{p/2})}\mu(d\phi) \\
&\leq CE\left[\liminf_{k\to\infty} \int_0^1 \left\| X_t^{M_{N(k)},N(k),(2)} \right\|^2_{B_2^{1-\varepsilon}(\nu)} dt\right] + C.
\end{aligned}$$

Therefore, in view of Lemma 7.1, Theorem 1.2(iii) holds. □

7.4. Non-Gaussianity of the limit measure

In this section we prove the non-Gaussianity of the Φ^4_3-measure obtained in Theorem 1.1 in a similar way to the proof of non-Gaussianity in [**GH21**]. Unlike the Φ^4_3-measure obtained in [**GH21**], we do not have such a nice integrability of our Φ^4_3-measure (see [**GH21**, Proposition 4.11]). For this reason we modify the proof in [**GH21**] and prove it by applying our Lemma 7.1. Similarly to Section 7.3, we choose α, γ, η, ε and q as in Theorem 6.10. Moreover, by applying the Skorohod representation theorem we assume that $X^{M_{N(k)},N(k)}$ converges to X in $C([0,\infty); B_{4/3}^{-1/2-\varepsilon}(\nu) \cap B_{12/5}^{-1/2-\varepsilon}(\nu^{9/5}))$ almost surely.

LEMMA 7.2. *For $f \in \mathcal{S}(\mathbb{R}^3)$ and $t \in (-\infty, \infty)$, we have*

$$\begin{aligned}
&\lim_{M,N\to\infty} E\left[\left\langle f, \mathcal{Z}_t^{(1,N)} \right\rangle^3 \left\langle f, P_{M,N} \mathcal{Z}_t^{(0,3,M,N)} \right\rangle\right] \\
&= \int_{\mathbb{R}^3} \int_0^\infty \left(e^{s(\triangle - m_0^2)} \left[2(m_0^2 - \triangle)\right]^{-1} f\right)^3 (x) \left(e^{s(\triangle - m_0^2)} f\right)(x) ds dx.
\end{aligned}$$

PROOF. Since for any $t \in (-\infty, \infty)$

$$\begin{aligned}
&E\left[\left\langle f, \mathcal{Z}_t^{(1,N)} \right\rangle^3 \left\langle f, P_{M,N} \mathcal{Z}_t^{(0,3,M,N)} \right\rangle\right] \\
&= \int_{-\infty}^t E\left[\langle P_N Z_t, f\rangle^3 \left\langle \mathcal{Z}_s^{(3,N)}, P_{M,N} e^{(t-s)(\triangle - m_0^2)} P_{M,N}^* f\right\rangle\right] ds,
\end{aligned}$$

by the product formula (see [**NP12**, Theorem 2.7.10]) or by the calculation of the expectation of Gaussian polynomials by pairing (see [**Jan97**, Theorem 1.28]) we have

$$\begin{aligned}
&E\left[\left\langle f, \mathcal{Z}_t^{(1,N)} \right\rangle^3 \left\langle f, P_{M,N} \mathcal{Z}_t^{(0,3,M,N)} \right\rangle\right] \\
&= \int_{-\infty}^t \int_{\mathbb{R}^3} E\left[\langle f, P_N Z_t\rangle \langle \delta_x, P_N Z_s\rangle\right]^3 \left(P_{M,N} e^{(t-s)(\triangle - m_0^2)} P_{M,N}^* f\right)(x) dx ds.
\end{aligned}$$

Hence, Lemma 3.1 implies

$$
\begin{aligned}
&E\left[\left\langle f, \mathcal{Z}_t^{(1,N)}\right\rangle^3 \left\langle f, \mathcal{Z}_t^{(0,3,M,N)}\right\rangle\right] \\
&= \int_{-\infty}^t \int_{\mathbb{R}^3} \langle V_N(t-s)f, \delta_x\rangle^3 \left(P_{M,N} e^{(t-s)(\triangle - m_0^2)} P_{M,N}^* f\right)(x) dx ds \\
&= \int_{-\infty}^t \int_{\mathbb{R}^3} \left(V_N(t-s)f\right)^3(x) \left(P_{M,N} e^{(t-s)(\triangle - m_0^2)} P_{M,N}^* f\right)(x) dx ds.
\end{aligned}
$$

By taking limit as $M, N \to \infty$ and exploiting that $f \in \mathcal{S}(\mathbb{R}^3)$, we get the assertion. □

LEMMA 7.3. *For $j \in \mathbb{N} \cup \{-1, 0\}$ and $y \in \mathbb{R}^3$, it holds that*

$$
\int_{\mathbb{R}^3} \int_0^\infty \left(e^{t(\triangle - m_0^2)} \left[2(m_0^2 - \triangle)\right]^{-1} \Delta_j \delta_y\right)^3 (x) \left(e^{t(\triangle - m_0^2)} \Delta_j \delta_y\right)(x) dt dx \asymp 2^j,
$$

where $a_j \asymp b_j$ means that there exists positive constants c_1 and c_2 satisfying $c_1 a_j \leq b_j \leq c_2 a_j$ for all $j \in \mathbb{N} \cup \{-1, 0\}$. Moreover, the integration in the left-hand side is independent of y.

PROOF. For $x \in \mathbb{R}^3$, we have

$$
\begin{aligned}
&\int_0^\infty \left(e^{t(\triangle - m_0^2)} \left[2(m_0^2 - \triangle)\right]^{-1} \Delta_j \delta_y\right)^3 (x) \left(e^{t(\triangle - m_0^2)} \Delta_j \delta_y\right)(x) dt \\
&= \int_0^\infty \int_{\mathbb{R}^3} \int_{\mathbb{R}^3} \int_{\mathbb{R}^3} \int_{\mathbb{R}^3} \left(\prod_{i=1}^3 \frac{e^{-t(m_0^2 + |\xi_i|^2)} \psi_j(\xi_i)}{2(m_0^2 + |\xi_i|^2)} (\mathcal{F}\delta_y)(\xi_i)\right) \\
&\quad \times \left(e^{-t(m_0^2 + |\xi_4|^2)} \psi_j(\xi_4) (\mathcal{F}\delta_y)(\xi_4)\right) (2\pi)^{-6} e^{\sqrt{-1}(\xi_1 + \xi_2 + \xi_3 + \xi_4)\cdot x} d\xi_1 d\xi_2 d\xi_3 d\xi_4 dt \\
&= \frac{1}{8(2\pi)^{12}} \int_{\mathbb{R}^3} \int_{\mathbb{R}^3} \int_{\mathbb{R}^3} \int_{\mathbb{R}^3} \frac{\psi_j(\xi_1)\psi_j(\xi_2)\psi_j(\xi_3)\psi_j(\xi_4)}{(m_0^2 + |\xi_1|^2)(m_0^2 + |\xi_2|^2)(m_0^2 + |\xi_3|^2)} \\
&\quad \times \frac{e^{-\sqrt{-1}(\xi_1 + \xi_2 + \xi_3 + \xi_4)\cdot y}}{4m_0^2 + |\xi_1|^2 + |\xi_2|^2 + |\xi_3|^2 + |\xi_4|^2} e^{\sqrt{-1}(\xi_1 + \xi_2 + \xi_3 + \xi_4)\cdot x} d\xi_1 d\xi_2 d\xi_3 d\xi_4.
\end{aligned}
$$

Hence, by changing variables in the integrations we see that

$$
\begin{aligned}
&\int_{\mathbb{R}^3} \int_0^\infty \left(e^{t(\triangle - m_0^2)} \left[2(m_0^2 - \triangle)\right]^{-1} \Delta_j \delta_y\right)^3 (x) \left(e^{t(\triangle - m_0^2)} \Delta_j \delta_y\right)(x) dt dx \\
&= \frac{2^j}{8(2\pi)^{12}} \int_{\mathbb{R}^3} \left(\int_{\mathbb{R}^3} \int_{\mathbb{R}^3} \int_{\mathbb{R}^3} \int_{\mathbb{R}^3} \frac{\psi(\xi_1)\psi(\xi_2)\psi(\xi_3)\psi(\xi_4)}{(2^{-2j}m_0^2 + |\xi_1|^2)(2^{-2j}m_0^2 + |\xi_2|^2)(2^{-2j}m_0^2 + |\xi_3|^2)}\right. \\
&\quad \left. \times \frac{e^{-\sqrt{-1}(\xi_1 + \xi_2 + \xi_3 + \xi_4)\cdot(2^j y)}}{2^{2-2j}m_0^2 + |\xi_1|^2 + |\xi_2|^2 + |\xi_3|^2 + |\xi_4|^2} e^{\sqrt{-1}(\xi_1 + \xi_2 + \xi_3 + \xi_4)\cdot x} d\xi_1 d\xi_2 d\xi_3 d\xi_4\right) dx.
\end{aligned}
$$

Since

$$\begin{aligned}
&\int_{\mathbb{R}^3}\left(\int_{\mathbb{R}^3}\int_{\mathbb{R}^3}\int_{\mathbb{R}^3}\int_{\mathbb{R}^3}\frac{\psi(\xi_1)\psi(\xi_2)\psi(\xi_3)\psi(\xi_4)}{(2^{-2j}m_0^2+|\xi_1|^2)(2^{-2j}m_0^2+|\xi_2|^2)(2^{-2j}m_0^2+|\xi_3|^2)}\right.\\
&\quad\left.\times\frac{e^{-\sqrt{-1}(\xi_1+\xi_2+\xi_3+\xi_4)\cdot(2^jy)}}{2^{2-2j}m_0^2+|\xi_1|^2+|\xi_2|^2+|\xi_3|^2+|\xi_4|^2}e^{\sqrt{-1}(\xi_1+\xi_2+\xi_3+\xi_4)\cdot x}d\xi_1d\xi_2d\xi_3d\xi_4\right)dx\\
&=\lim_{\varepsilon\downarrow 0}\int_{\mathbb{R}^3}\left(\int_{\mathbb{R}^3}\int_{\mathbb{R}^3}\int_{\mathbb{R}^3}\int_{\mathbb{R}^3}\frac{\psi(\xi_1)\psi(\xi_2)\psi(\xi_3)\psi(\xi_4)}{(2^{-2j}m_0^2+|\xi_1|^2)(2^{-2j}m_0^2+|\xi_2|^2)(2^{-2j}m_0^2+|\xi_3|^2)}\right.\\
&\quad\times\frac{e^{-\sqrt{-1}(\xi_1+\xi_2+\xi_3+\xi_4)\cdot(2^jy)}}{2^{2-2j}m_0^2+|\xi_1|^2+|\xi_2|^2+|\xi_3|^2+|\xi_4|^2}\\
&\quad\left.\times e^{-\varepsilon|x|^2}e^{\sqrt{-1}(\xi_1+\xi_2+\xi_3+\xi_4)\cdot x}d\xi_1d\xi_2d\xi_3d\xi_4\right)dx\\
&=\lim_{\varepsilon\downarrow 0}\left(\frac{\pi}{\varepsilon}\right)^{3/2}\int_{\mathbb{R}^3}\int_{\mathbb{R}^3}\int_{\mathbb{R}^3}\int_{\mathbb{R}^3}\frac{\psi(\xi_1)\psi(\xi_2)\psi(\xi_3)\psi(\xi_4)}{(2^{-2j}m_0^2+|\xi_1|^2)(2^{-2j}m_0^2+|\xi_2|^2)(2^{-2j}m_0^2+|\xi_3|^2)}\\
&\quad\times\frac{e^{-\sqrt{-1}(\xi_1+\xi_2+\xi_3+\xi_4)\cdot(2^jy)}}{2^{2-2j}m_0^2+|\xi_1|^2+|\xi_2|^2+|\xi_3|^2+|\xi_4|^2}e^{-|\xi_1+\xi_2+\xi_3+\xi_4|^2/4\varepsilon}d\xi_1d\xi_2d\xi_3d\xi_4\\
&=(2\pi)^3\int_{\mathbb{R}^3}\int_{\mathbb{R}^3}\int_{\mathbb{R}^3}\int_{\mathbb{R}^3}\frac{\psi(\xi_1)\psi(\xi_2)\psi(\xi_3)\psi(\xi_4)}{(2^{-2j}m_0^2+|\xi_1|^2)(2^{-2j}m_0^2+|\xi_2|^2)(2^{-2j}m_0^2+|\xi_3|^2)}\\
&\quad\times\frac{e^{-\sqrt{-1}(\xi_1+\xi_2+\xi_3+\xi_4)\cdot(2^jy)}}{2^{2-2j}m_0^2+|\xi_1|^2+|\xi_2|^2+|\xi_3|^2+|\xi_4|^2}\delta_0(\xi_1+\xi_2+\xi_3+\xi_4)d\xi_1d\xi_2d\xi_3d\xi_4,
\end{aligned}$$

we get the assertion. □

Now we are ready to prove Theorem 1.2(iv).

Proof of Theorem 1.2(iv). By assuming ad absurdum that μ is a Gaussian measure on $\mathcal{S}'(\mathbb{R}^3)$, we show that we are lead to a contradiction. Let $j\in\mathbb{N}\cup\{-1,0\}$ and $t\in(0,\infty)$. We remark that the law of X_t is μ for $t\in[0,\infty)$ and that the Gaussianity assumption implies that for any j, $\langle\Delta_j\delta_x,X_t\rangle$ has all moments with respect to the probability measure P for $x\in\mathbb{R}^3$. Hence, $\langle\Delta_j\delta_x,X_t-Z_t\rangle$ also has all moments with respect to the probability measure P for $x\in\mathbb{R}^3$. Since the laws of $\langle\Delta_j\delta_x,X_t\rangle$ and $\langle\Delta_j\delta_x,Z_t\rangle$ are Gaussian, we get

(7.6)
$$E\left[\langle\Delta_j\delta_x,X_t\rangle^4\right]=3E\left[\langle\Delta_j\delta_x,X_t\rangle^2\right]^2,\quad E\left[\langle\Delta_j\delta_x,Z_t\rangle^4\right]=3E\left[\langle\Delta_j\delta_x,Z_t\rangle^2\right]^2.$$

Hence, we have

(7.7)
$$\begin{aligned}
&E\left[\langle\Delta_j\delta_x,X_t-Z_t\rangle^4\right]+4E\left[\langle\Delta_j\delta_x,X_t-Z_t\rangle^3\langle\Delta_j\delta_x,Z_t\rangle\right]\\
&\quad+6E\left[\langle\Delta_j\delta_x,X_t-Z_t\rangle^2\langle\Delta_j\delta_x,Z_t\rangle^2\right]+4E\left[\langle\Delta_j\delta_x,X_t-Z_t\rangle\langle\Delta_j\delta_x,Z_t\rangle^3\right]\\
&=3E\left[\langle\Delta_j\delta_x,X_t-Z_t\rangle^2\right]^2+12E\left[\langle\Delta_j\delta_x,X_t-Z_t\rangle\langle\Delta_j\delta_x,Z_t\rangle\right]^2\\
&\quad+12E\left[\langle\Delta_j\delta_x,X_t-Z_t\rangle^2\right]E\left[\langle\Delta_j\delta_x,X_t-Z_t\rangle\langle\Delta_j\delta_x,Z_t\rangle\right]\\
&\quad+6\left(E\left[\langle\Delta_j\delta_x,X_t-Z_t\rangle^2\right]+2E\left[\langle\Delta_j\delta_x,X_t-Z_t\rangle\langle\Delta_j\delta_x,Z_t\rangle\right]\right)E\left[\langle\Delta_j\delta_x,Z_t\rangle^2\right]
\end{aligned}$$

for all $t \in [0, \infty)$. This inequality, Hölder's inequality and (7.6) imply

$$\begin{aligned}
&\left|E\left[\langle \Delta_j \delta_x, X_t - Z_t\rangle \langle \Delta_j \delta_x, Z_t\rangle^3\right]\right| \\
&\leq CE\left[\langle \Delta_j \delta_x, X_t - Z_t\rangle^2\right] E\left[\langle \Delta_j \delta_x, Z_t\rangle^2\right] \\
&\quad + C\sum_{l=3}^{4} E\left[\langle \Delta_j \delta_x, X_t - Z_t\rangle^4\right]^{l/4} E\left[\langle \Delta_j \delta_x, Z_t\rangle^2\right]^{(4-l)/2} \\
&\quad + \left|E\left[\langle \Delta_j \delta_x, X_t - Z_t\rangle \langle \Delta_j \delta_x, Z_t\rangle\right]\right| E\left[\langle \Delta_j \delta_x, Z_t\rangle^2\right].
\end{aligned}$$

On the other hand,

$$\begin{aligned}
&\int_{\mathbb{R}^3} E\left[\langle \Delta_j \delta_x, X_t - Z_t\rangle \langle \Delta_j \delta_x, Z_t\rangle^3\right] \nu^3(x) dx \\
&= \int_{\mathbb{R}^3} E\left[(\Delta_j (X_t - Z_t))(x) (\Delta_j Z_t)^3(x)\right] \nu^3(x) dx \\
&= E\left[\int_{\mathbb{R}^3} [\Delta_j (X_t - Z_t)] (\Delta_j Z_t)^3 \nu^3 dx\right].
\end{aligned}$$

Note that the translation invariance of the law of Z_t implies the independence of $E\left[\langle \Delta_j \delta_x, Z_t\rangle^2\right]$ in $x \in \mathbb{R}^3$, and that other terms in (7.7) are calculated similarly. Hence, we have

$$\begin{aligned}
&\left|E\left[\int_{\mathbb{R}^3} [\Delta_j (X_t - Z_t)] (\Delta_j Z_t)^3 \nu^3 dx\right]\right| \\
&\leq CE\left[\|\Delta_j (X_t - Z_t)\|_{L^2(\nu^3)}^2\right] E\left[\|\Delta_j Z_t\|_{L^2(\nu^3)}^2\right] \\
&\quad + C\sum_{l=3}^{4} E\left[\|\Delta_j (X_t - Z_t)\|_{L^4(\nu^3)}^4\right]^{l/4} E\left[\|\Delta_j Z_t\|_{L^2(\nu^3)}^2\right]^{(4-l)/2} \\
&\quad + C\left|E\left[\int_{\mathbb{R}^3} [\Delta_j (X_t - Z_t)] (\Delta_j Z_t) \nu^3 dx\right]\right| E\left[\|\Delta_j Z_t\|_{L^2(\nu^3)}^2\right].
\end{aligned} \tag{7.8}$$

Since

$$E\left[\|\Delta_j Z_t\|_{L^2(\nu^3)}^2\right] \leq 2^{(1+2\varepsilon)j} E\left[\|Z_t\|_{B_2^{-1/2-\varepsilon}(\nu^3)}^2\right],$$

by Proposition 3.3 we have

$$E\left[\|\Delta_j Z_t\|_{L^2(\nu^3)}^2\right] \leq C 2^{(1+2\varepsilon)j}. \tag{7.9}$$

In view of the smoothing property of Δ_j and Proposition 2.4, it holds that

$$\begin{aligned}
&\limsup_{k\to\infty} \sup_{s\in[0,T]} \left\|\Delta_j \left(X_s^{M_{N(k)}, N(k)} - Z_s\right) - \Delta_j (X_s - Z_s)\right\|_{L^4(\nu^3)} \\
&\leq C_j \limsup_{k\to\infty} \sup_{s\in[0,T]} \left\|\Delta_j \left(X_s^{M_{N(k)}, N(k)} - Z_s\right) - \Delta_j (X_s - Z_s)\right\|_{B_{4/3}^{-1/2-\varepsilon}(\nu)} \\
&= 0.
\end{aligned}$$

From this fact, stationarity of (X_t, Z_t) and Fatou's lemma we have

$$\begin{aligned} E\left[\|\Delta_j(X_t - Z_t)\|^4_{L^4(\nu^3)}\right] &= \int_0^1 E\left[\|\Delta_j(X_s - Z_s)\|^4_{L^4(\nu^3)}\right] ds \\ &= \int_0^1 E\left[\liminf_{k\to\infty}\left\|\Delta_j\left(X_s^{M_{N(k)},N(k)} - Z_s\right)\right\|^4_{L^4(\nu^3)}\right] ds \\ &\le E\left[\liminf_{k\to\infty}\int_0^1 \left\|X_s^{M_{N(k)},N(k)} - Z_s\right\|^4_{L^4(\nu^3)} ds\right] \end{aligned}$$

Similarly

$$\begin{aligned} &E\left[\|\Delta_j(X_t - Z_t)\|^2_{L^2(\nu^3)}\right] \\ &\le 2^{-[(1/2)-\varepsilon]j} E\left[\liminf_{k\to\infty}\int_0^1 \left\|X_s^{M_{N(k)},N(k)} - Z_s\right\|^2_{B_2^{1/2-\varepsilon}(\nu^3)} ds\right]. \end{aligned}$$

From (7.8), (7.9), these inequalities and Lemma 7.1 we have

$$\begin{aligned} &\left|E\left[\int_{\mathbb{R}^3}[\Delta_j(X_t - Z_t)](\Delta_j Z_t)^3\nu^3 dx\right]\right| \\ &\le C2^{(1+2\varepsilon)j}\left|E\left[\int_{\mathbb{R}^3}[\Delta_j(X_t - Z_t)](\Delta_j Z_t)\nu^3 dx\right]\right| + C2^{[(1/2)+3\varepsilon]j}. \end{aligned} \tag{7.10}$$

Almost-sure convergence of $\{(X^{M_{N(k)},N(k)}, \mathcal{Z}^{(1,M_{N(k)},N(k))}, \mathcal{Z}^{(0,3,M_{N(k)},N(k))})\}$ and Fatou's lemma imply

$$\begin{aligned} &E\left[\int_{\mathbb{R}^3}[\Delta_j(X_t - Z_t)](\Delta_j Z_t)\nu^3 dx\right] \\ &= \int_0^1 E\left[\lim_{k\to\infty}\left(\int_{\mathbb{R}^3}\left(\Delta_j X_s^{M_{N(k)},N(k),(2)}\right)\left(\Delta_j \mathcal{Z}_s^{(1,M_{N(k)},N(k))}\right)\nu^3 dx\right.\right. \\ &\quad \left.\left. -\lambda\int_{\mathbb{R}^3}\left(\Delta_j P_{M_{N(k)},N(k)}\mathcal{Z}_s^{(0,3,M_{N(k)},N(k))}\right)\left(\Delta_j \mathcal{Z}_s^{(1,M_{N(k)},N(k))}\right)\nu^3 dx\right)\right] ds \\ &\le E\left[\liminf_{k\to\infty}\int_0^1 \left\|\Delta_j X_s^{M_{N(k)},N(k),(2)}\right\|_{L^2(\nu^3)}\left\|\Delta_j \mathcal{Z}_s^{(1,M_{N(k)},N(k))}\right\|_{L^2(\nu^3)} ds\right] \\ &\quad -\lambda\int_0^1 \lim_{k\to\infty} E\left[\int_{\mathbb{R}^3}\left(\Delta_j P_{M_{N(k)},N(k)}\mathcal{Z}_s^{(0,3,M_{N(k)},N(k))}\right)\right. \\ &\qquad\qquad \left.\times\left(\Delta_j \mathcal{Z}_s^{(1,M_{N(k)},N(k))}\right)\nu^3 dx\right] ds. \end{aligned}$$

The orthogonality of Wick polynomials yields

$$E\left[\left(\Delta_j P_{M_{N(k)},N(k)}\mathcal{Z}_s^{(0,3,M_{N(k)},N(k))}\right)(x)\left(\Delta_j \mathcal{Z}_s^{(1,M_{N(k)},N(k))}\right)(x)\right] = 0$$

for $x \in \mathbb{R}^3$ and $s \in [0,\infty)$, so that we have for some positive constant C:

$$\begin{aligned} &E\left[\int_{\mathbb{R}^3}[\Delta_j(X_t - Z_t)](\Delta_j Z_t)\nu^3 dx\right] \\ &\le C2^{-[(1/2)-2\varepsilon]j} \\ &\quad\times E\left[\liminf_{k\to\infty}\int_0^1 \left\|X_s^{M_{N(k)},N(k),(2)}\right\|_{B_2^{1-\varepsilon}(\nu^3)}\left\|\mathcal{Z}_s^{(1,M_{N(k)},N(k))}\right\|_{B_2^{-(1/2)-\varepsilon}(\nu^3)} ds\right]. \end{aligned}$$

Hence, in view of Proposition 3.3 and Lemma 7.1, from (7.10) we have

$$\left| E\left[\int_{\mathbb{R}^3} [\Delta_j (X_t - Z_t)](\Delta_j Z_t)^3 \nu^3 dx \right] \right| \le C 2^{[(1/2)+4\varepsilon]j} \tag{7.11}$$

for some positive C. For the left-hand side of this inequality deduce it as

$$\begin{aligned}
&\lambda \left| E\left[\int_0^1 \lim_{k\to\infty} \int_{\mathbb{R}^3} \left(\Delta_j P_{M_{N(k)},N(k)} \mathcal{Z}_s^{(0,3,M_{N(k)},N(k))} \right) \right.\right. \\
&\qquad\qquad \left.\left. \times \left(\Delta_j \mathcal{Z}_s^{(1,M_{N(k)},N(k))} \right)^3 \nu^3 dx ds \right] \right| \\
&\le \left| E\left[\int_{\mathbb{R}^3} [\Delta_j (X_t - Z_t)](\Delta_j Z_t)^3 \nu^3 dx \right] \right| \\
&\quad + \left| E\left[\int_0^1 \lim_{k\to\infty} \int_{\mathbb{R}^3} \left(\Delta_j X_s^{M_{N(k)},N(k),(2)} \right) \left(\Delta_j \mathcal{Z}_s^{(1,M_{N(k)},N(k))} \right)^3 \nu^3 dx ds \right] \right| \\
&\le \left| E\left[\int_{\mathbb{R}^3} [\Delta_j (X_t - Z_t)](\Delta_j Z_t)^3 \nu^3 dx \right] \right| \\
&\quad + E\left[\liminf_{k\to\infty} \int_0^1 \left\| \Delta_j X_s^{M_{N(k)},N(k),(2)} \right\|_{L^2(\nu^3)} \left\| \Delta_j \mathcal{Z}_s^{(1,M_{N(k)},N(k))} \right\|_{L^6(\nu^3)}^3 ds \right].
\end{aligned}$$

Since Proposition 3.3 and Lemma 7.1 imply

$$\begin{aligned}
&E\left[\liminf_{k\to\infty} \int_0^1 \left\| \Delta_j X_s^{M_{N(k)},N(k),(2)} \right\|_{L^2(\nu^3)} \left\| \Delta_j \mathcal{Z}_s^{(1,M_{N(k)},N(k))} \right\|_{L^6(\nu^3)}^3 ds \right] \\
&\le C 2^{[(1/2)+4\varepsilon]j} \\
&\quad \times E\left[\liminf_{k\to\infty} \int_0^1 \left\| X_s^{M_{N(k)},N(k),(2)} \right\|_{B_2^{1-\varepsilon}(\nu^3)} \left\| \mathcal{Z}_s^{(1,M_{N(k)},N(k))} \right\|_{B_6^{-(1/2)-\varepsilon}(\nu^3)}^3 ds \right] \\
&\le C 2^{[(1/2)+4\varepsilon]j},
\end{aligned}$$

by applying these inequalities to (7.11) we have

$$\begin{aligned}
&\lambda \left| E\left[\int_0^1 \lim_{k\to\infty} \int_{\mathbb{R}^3} \left(\Delta_j P_{M_{N(k)},N(k)} \mathcal{Z}_s^{(0,3,M_{N(k)},N(k))} \right) \right.\right. \\
&\qquad\qquad \left.\left. \times \left(\Delta_j \mathcal{Z}_s^{(1,M_{N(k)},N(k))} \right)^3 \nu^3 dx ds \right] \right| \\
&\le C 2^{[(1/2)+4\varepsilon]j}.
\end{aligned} \tag{7.12}$$

In view of Proposition 3.3, $\Delta_j \mathcal{Z}_s^{(0,3,M_{N(k)},N(k))}$ and $\Delta_j \mathcal{Z}_s^{(1,M_{N(k)},N(k))}$ have nice integrability with respect to both s and the probability measure P. Hence, by Lemmas 7.2 and 7.3 we have

$$\begin{aligned}
&E\left[\int_0^1 \lim_{k\to\infty} \int_{\mathbb{R}^3} \left(\Delta_j P_{M_{N(k)},N(k)} \mathcal{Z}_s^{(0,3,M_{N(k)},N(k))} \right) \left(\Delta_j \mathcal{Z}_s^{(1,M_{N(k)},N(k))} \right)^3 \nu^3 dx ds \right] \\
&= \lim_{k\to\infty} \int_0^1 \int_{\mathbb{R}^3} E\left[\left\langle \Delta_j \delta_x, P_{M_{N(k)},N(k)} \mathcal{Z}_s^{(0,3,M_{N(k)},N(k))} \right\rangle \left\langle \Delta_j \delta_x, \mathcal{Z}_s^{(1,M_{N(k)},N(k))} \right\rangle^3 \right] \\
&\quad \times \nu^3(x) dx ds \\
&\ge c 2^j
\end{aligned}$$

where c is a positive constant. This fact conflicts with (7.12) and hence Theorem 1.2(iv) is proven. □

APPENDIX A

A sufficient condition for integrability

PROPOSITION A.1. Let X be a nonnegative random variable, $\theta \in (0,1)$ and let c_1, c_2 be nonnegative constants. If for sufficiently large $n \in \mathbb{N}$ it holds that

$$E\left[\min\{X,n\}\right] \leq c_1 E\left[\min\{X^\theta, n\}\right] + c_2,$$

then

$$E[X] \leq \frac{1}{1-\theta}\left(c_1^{1/(1-\theta)} + c_2\right).$$

In particular, X is integrable with respect to the probability measure.

PROOF. It is sufficient to prove the case where $c_1 > 0$. Since

$$\begin{aligned} E\left[\min\{X^\theta, n\}\right] &= \int_0^\infty P\left(\min\{X^\theta, n\} > \lambda\right) d\lambda \\ &\leq \int_0^n P\left(X > \lambda^{1/\theta}\right) d\lambda, \end{aligned}$$

by changing variables we have

$$\begin{aligned} c_1 E\left[\min\{X^\theta, n\}\right] &\leq \theta c_1 \int_0^{n^\theta} P(X > \eta)\, \eta^{-(1-\theta)} d\eta \\ &\leq \theta \int_{c_1^{1/(1-\theta)}}^{n^\theta} P(X > \eta)\, d\eta + \theta c_1 \int_0^{c_1^{1/(1-\theta)}} \eta^{-(1-\theta)} d\eta. \end{aligned}$$

Hence, we have

$$c_1 E\left[\min\{X^\theta, n\}\right] \leq \theta E\left[\min\{X,n\}\right] + c_1^{1/(1-\theta)}.$$

This inequality and the assumption yields

$$E\left[\min\{X,n\}\right] \leq \frac{1}{1-\theta}\left(c_1^{1/(1-\theta)} + c_2\right).$$

By taking the limit as $n \to \infty$ we have the assertion. □

Bibliography

[ABDVG22] S. Albeverio, L. Borasi, F. C. De Vecchi, and M. Gubinelli, *Grassmannian stochastic analysis and the stochastic quantization of Euclidean fermions*, Probab. Theory Related Fields **183** (2022), no. 3-4, 909–995, DOI 10.1007/s00440-022-01136-x. MR4453319

[ABR99] S. Albeverio, V. Bogachev, and M. Röckner, *On uniqueness of invariant measures for finite- and infinite-dimensional diffusions*, Comm. Pure Appl. Math. **52** (1999), no. 3, 325–362, DOI 10.1002/(SICI)1097-0312(199903)52:3<325::AID-CPA2>3.0.CO;2-V. MR1656067

[ADVG20] S. Albeverio, F. C. De Vecchi, and M. Gubinelli, *Elliptic stochastic quantization*, Ann. Probab. **48** (2020), no. 4, 1693–1741, DOI 10.1214/19-AOP1404. MR4124523

[ADVG21] S. Albeverio, F. C. De Vecchi, and M. Gubinelli, *The elliptic stochastic quantization of some two dimensional Euclidean QFTs* (English, with English and French summaries), Ann. Inst. Henri Poincaré Probab. Stat. **57** (2021), no. 4, 2372–2414, DOI 10.1214/20-aihp1145. MR4415393

[AFHKL86] S. Albeverio, R. Høegh-Krohn, J. E. Fenstad, and T. Lindstrøm, *Nonstandard methods in stochastic analysis and mathematical physics*, Pure and Applied Mathematics, vol. 122, Academic Press, Inc., Orlando, FL, 1986. MR859372

[AGHK79] S. Albeverio, G. Gallavotti, and R. Høegh-Krohn, *Some results for the exponential interaction in two or more dimensions*, Comm. Math. Phys. **70** (1979), no. 2, 187–192. MR553970

[AGR01] S. Albeverio, R. Gielerak, and F. Russo, *On the paths Hölder continuity in models of Euclidean quantum field theory*, Stochastic Anal. Appl. **19** (2001), no. 5, 677–702, DOI 10.1081/SAP-120000217. MR1857892

[AHK73] S. Albeverio and R. Høegh-Krohn, *Uniqueness of the physical vacuum and the Wightman functions in the infinite volume limit for some non polynomial interactions*, Comm. Math. Phys. **30** (1973), 171–200. MR345553

[AHK74] S. Albeverio and R. Høegh-Krohn, *The Wightman axioms and the mass gap for strong interactions of exponential type in two-dimensional space-time*, J. Functional Analysis **16** (1974), 39–82, DOI 10.1016/0022-1236(74)90070-6. MR356761

[AHK77] S. Albeverio and R. Høegh-Krohn, *Dirichlet forms and diffusion processes on rigged Hilbert spaces*, Z. Wahrscheinlichkeitstheorie und Verw. Gebiete **40** (1977), no. 1, 1–57, DOI 10.1007/BF00535706. MR455133

[AHKS77] S. Albeverio, R. Høegh-Krohn, and L. Streit, *Energy forms, Hamiltonians, and distorted Brownian paths*, J. Mathematical Phys. **18** (1977), no. 5, 907–917, DOI 10.1063/1.523359. MR446236

[AHKZ89] S. Albeverio, R. Høegh-Krohn, and B. Zegarliński, *Uniqueness and global Markov property for Euclidean fields: the case of general polynomial interactions*, Comm. Math. Phys. **123** (1989), no. 3, 377–424. MR1003426

[AHP$^+$89a] S. Albeverio, T. Hida, J. Potthoff, M. Röckner, and L. Streit, *Dirichlet forms in terms of white noise analysis. I. Construction and QFT examples*, Rev. Math. Phys. **1** (1989), no. 2-3, 291–312, DOI 10.1142/S0129055X89000134. MR1070092

[AHP$^+$89b] S. Albeverio, T. Hida, J. Potthoff, M. Röckner, and L. Streit, *Dirichlet forms in terms of white noise analysis. II. Closability and diffusion processes*, Rev. Math. Phys. **1** (1989), no. 2-3, 313–323, DOI 10.1142/S0129055X89000146. MR1070093

[AHR96] S. Albeverio, Z. Haba, and F. Russo, *Trivial solutions for a non-linear two-space-dimensional wave equation perturbed by space-time white noise*, Stochastics Stochastics Rep. **56** (1996), no. 1-2, 127–160, DOI 10.1080/17442509608834039. MR1396758

[AHR01] S. Albeverio, Z. Haba, and F. Russo, *A two-space dimensional semilinear heat equation perturbed by (Gaussian) white noise*, Probab. Theory Related Fields **121** (2001), no. 3, 319–366, DOI 10.1007/s004400100153. MR1867426

[AK20] S. Albeverio and S. Kusuoka, *The invariant measure and the flow associated to the Φ^4_3-quantum field model*, Ann. Sc. Norm. Super. Pisa Cl. Sci. (5) **20** (2020), no. 4, 1359–1427. MR4201185

[AKKR09] S. Albeverio, Y. Kondratiev, Y. Kozitsky, and M. Röckner, *The statistical mechanics of quantum lattice systems*, EMS Tracts in Mathematics, vol. 8, European Mathematical Society (EMS), Zürich, 2009. A path integral approach, DOI 10.4171/070. MR2548038

[AKMR23] S. Albeverio, H. Kawabi, S.-R. Mihalache, and M. Röckner, *Strong uniqueness for Dirichlet operators related to stochastic quantization under exponential/trigonometric interactions on the two-dimensional torus*, Ann. Sc. Norm. Super. Pisa Cl. Sci. (5) **24** (2023), no. 1, 33–69, DOI 10.2422/2036-2145.202105_106. MR4587740

[AKR97] S. Albeverio, Y. G. Kondratiev, and M. Röckner, *Ergodicity for the stochastic dynamics of quasi-invariant measures with applications to Gibbs states*, J. Funct. Anal. **149** (1997), no. 2, 415–469, DOI 10.1006/jfan.1997.3099. MR1472365

[AKR12] S. Albeverio, H. Kawabi, and M. Röckner, *Strong uniqueness for both Dirichlet operators and stochastic dynamics to Gibbs measures on a path space with exponential interactions*, J. Funct. Anal. **262** (2012), no. 2, 602–638, DOI 10.1016/j.jfa.2011.09.023. MR2854715

[AKYY21] S. Albeverio, T. Kagawa, Y. Yahagi, and M. W. Yoshida, *Non-local Markovian symmetric forms on infinite dimensional spaces I. The closability and quasi-regularity*, Comm. Math. Phys. **388** (2021), no. 2, 659–706, DOI 10.1007/s00220-021-04186-9. MR4334244

[AL04] S. Albeverio and S. Liang, *A remark on different lattice approximations and continuum limits for ϕ^4_2-fields*, Random Oper. Stochastic Equations **12** (2004), no. 4, 313–318, DOI 10.1163/1569397042722346. MR2108186

[AL07] S. Albeverio and S. Liang, *A note on the renormalized square of free quantum fields in space-time dimension $d \geq 4$*, Bull. Sci. Math. **131** (2007), no. 1, 1–11, DOI 10.1016/j.bulsci.2006.02.004. MR2285582

[Alb03] S. Albeverio, *Theory of Dirichlet forms and applications*, Lectures on probability theory and statistics (Saint-Flour, 2000), Lecture Notes in Math., vol. 1816, Springer, Berlin, 2003, pp. 1–106, DOI 10.1007/3-540-44922-1. MR2009816

[ALZ06] S. Albeverio, S. Liang, and B. Zegarlinski, *Remark on the integration by parts formula for the ϕ^4_3-quantum field model*, Infin. Dimens. Anal. Quantum Probab. Relat. Top. **9** (2006), no. 1, 149–154, DOI 10.1142/S0219025706002275. MR2214506

[AM15] S. Albeverio and S. Mazzucchi, *An introduction to infinite-dimensional oscillatory and probabilistic integrals*, Stochastic analysis: a series of lectures, Progr. Probab., vol. 68, Birkhäuser/Springer, Basel, 2015, pp. 1–54, DOI 10.1007/978-3-0348-0909-2_1. MR3558120

[AMR15] S. Albeverio, Z.-M. Ma, and M. Röckner, *Quasi regular Dirichlet forms and the stochastic quantization problem*, Festschrift Masatoshi Fukushima, Interdiscip. Math. Sci., vol. 17, World Sci. Publ., Hackensack, NJ, 2015, pp. 27–58, DOI 10.1142/9789814596534_0003. MR3379334

[AR89] S. Albeverio and M. Röckner, *Classical Dirichlet forms on topological vector spaces—the construction of the associated diffusion process*, Probab. Theory Related Fields **83** (1989), no. 3, 405–434, DOI 10.1007/BF00964372. MR1017404

[AR90] S. Albeverio and M. Röckner, *Classical Dirichlet forms on topological vector spaces—closability and a Cameron-Martin formula*, J. Funct. Anal. **88** (1990), no. 2, 395–436, DOI 10.1016/0022-1236(90)90113-Y. MR1038449

[AR91] S. Albeverio and M. Röckner, *Stochastic differential equations in infinite dimensions: solutions via Dirichlet forms*, Probab. Theory Related Fields **89** (1991), no. 3, 347–386, DOI 10.1007/BF01198791. MR1113223

[ARY17] S. Albeverio, M. Röckner, and M. W. Yoshida, *Quantum fields*, Let us use white noise, World Sci. Publ., Hackensack, NJ, 2017, pp. 37–65. MR3588084

[AU15] S. Albeverio and S. Ugolini, *A Doob h-transform of the Gross-Pitaevskii Hamiltonian*, J. Stat. Phys. **161** (2015), no. 2, 486–508, DOI 10.1007/s10955-015-1337-3. MR3401027

[AY02] S. Albeverio and M. W. Yoshida, *H-C^1 maps and elliptic SPDEs with polynomial and exponential perturbations of Nelson's Euclidean free field*, J. Funct. Anal. **196** (2002), no. 2, 265–322, DOI 10.1016/S0022-1236(02)00008-3. MR1943094

[AY09] S. Albeverio and M. W. Yoshida, *Hida distribution construction of non-Gaussian reflection positive generalized random fields*, Infin. Dimens. Anal. Quantum Probab. Relat. Top. **12** (2009), no. 1, 21–49, DOI 10.1142/S0219025709003550. MR2509983

[Bał83] T. Bałaban, *Ultraviolet stability in field theory. The φ_3^4 model*, Scaling and self-similarity in physics (Bures-sur-Yvette, 1981/1982), Progr. Phys., vol. 7, Birkhäuser Boston, Boston, MA, 1983, pp. 297–319. MR733476

[Bat99] G. Battle, *Wavelets and renormalization*, Series in Approximations and Decompositions, vol. 10, World Scientific Publishing Co., Inc., River Edge, NJ, 1999, DOI 10.1142/3066. MR1688691

[BCD11] H. Bahouri, J.-Y. Chemin, and R. Danchin, *Fourier analysis and nonlinear partial differential equations*, Grundlehren der mathematischen Wissenschaften [Fundamental Principles of Mathematical Sciences], vol. 343, Springer, Heidelberg, 2011, DOI 10.1007/978-3-642-16830-7. MR2768550

[BCG+78] G. Benfatto, M. Cassandro, G. Gallavotti, F. Nicolò, E. Olivieri, E. Presutti, and E. Scacciatelli, *Some probabilistic techniques in field theory*, Comm. Math. Phys. **59** (1978), no. 2, 143–166. MR491611

[BCM88] V. S. Borkar, R. T. Chari, and S. K. Mitter, *Stochastic quantization of field theory in finite and infinite volume*, J. Funct. Anal. **81** (1988), no. 1, 184–206, DOI 10.1016/0022-1236(88)90117-6. MR967896

[BDH95] D. Brydges, J. Dimock, and T. R. Hurd, *The short distance behavior of $(\phi^4)_3$*, Comm. Math. Phys. **172** (1995), no. 1, 143–186. MR1346375

[BDV21] N. Barashkov and F. C. De Vecchi, *Elliptic stochastic quantization of Sinh-Gordon QFT*, Preprint, `arXiv:2108.12664`, (2021).

[BF84] A. Bovier and G. Felder, *Skeleton inequalities and the asymptotic nature of perturbation theory for Φ^4-theories in two and three dimensions*, Comm. Math. Phys. **93** (1984), no. 2, 259–275. MR742195

[BFS83] D. C. Brydges, J. Fröhlich, and A. D. Sokal, *A new proof of the existence and nontriviality of the continuum φ_2^4 and φ_3^4 quantum field theories*, Comm. Math. Phys. **91** (1983), no. 2, 141–186. MR723546

[BG20] N. Barashkov and M. Gubinelli, *A variational method for Φ_3^4*, Duke Math. J. **169** (2020), no. 17, 3339–3415, DOI 10.1215/00127094-2020-0029. MR4173157

[BG21] N. Barashkov and M. Gubinelli, *The Φ_3^4 measure via Girsanov's theorem*, Electron. J. Probab. **26** (2021), Paper No. 81, 29, DOI 10.1214/21-ejp635. MR4269211

[BGN82] G. Benfatto, G. Gallavotti, and F. Nicolò, *On the massive sine-Gordon equation in the first few regions of collapse*, Comm. Math. Phys. **83** (1982), no. 3, 387–410. MR649810

[BH20] I. Bailleul and M. Hoshino, *A tourist's guide to regularity structures*, To appear in EMS Surv. Math. Sci. **12** (2025), DOI 10.4171/EMSS/87.

[BK95a] Y. M. Berezansky and Y. G. Kondratiev, *Spectral methods in infinite-dimensional analysis. Vol. 1*, Mathematical Physics and Applied Mathematics, vol. 12/1, Kluwer Academic Publishers, Dordrecht, 1995. Translated from the 1988 Russian original by P. V. Malyshev and D. V. Malyshev and revised by the authors, DOI 10.1007/978-94-011-0509-5. MR1340626

[BK95b] Y. M. Berezansky and Y. G. Kondratiev, *Spectral methods in infinite-dimensional analysis. Vol. 2*, Mathematical Physics and Applied Mathematics, vol. 12/2, Kluwer Academic Publishers, Dordrecht, 1995. Translated from the 1988 Russian original by P. V. Malyshev and D. V. Malyshev and revised by the authors, DOI 10.1007/978-94-011-0509-5. MR1340627

[BL03] V. Betz and J. Lőrinczi, *Uniqueness of Gibbs measures relative to Brownian motion* (English, with English and French summaries), Ann. Inst. H. Poincaré Probab. Statist. **39** (2003), no. 5, 877–889, DOI 10.1016/S0246-0203(03)00021-9. MR1997216

[BLT75] N. N. Bogolubov, A. A. Logunov, and I. T. Todorov, *Introduction to axiomatic quantum field theory*, Mathematical Physics Monograph Series, No. 18, W. A. Benjamin, Inc., Reading, Mass.-London-Amsterdam, 1975. Translated from the Russian by Stephen A. Fulling and Ludmila G. Popova; Edited by Stephen A. Fulling. MR452277

[BSZ92] J. C. Baez, I. E. Segal, and Z.-F. Zhou, *Introduction to algebraic and constructive quantum field theory*, Princeton Series in Physics, Princeton University Press, Princeton, NJ, 1992, DOI 10.1515/9781400862504. MR1178936

[Bur] Charles Burnap, *The particle structure of boson quantum field theory models*, Ph.D Thesis, Harvard University, 1976.

[Bur77] C. Burnap, *Isolated one particle states in boson quantum field theory models*, Ann. Physics **104** (1977), no. 1, 184–196, DOI 10.1016/0003-4916(77)90050-1. MR443691

[CC18] R. Catellier and K. Chouk, *Paracontrolled distributions and the 3-dimensional stochastic quantization equation*, Ann. Probab. **46** (2018), no. 5, 2621–2679, DOI 10.1214/17-AOP1235. MR3846835

[CCHS22] A. Chandra, I. Chevyrev, M. Hairer, and H. Shen, *Stochastic quantisation of Yang-Mills-Higgs in 3D*, Preprint, arXiv:2201.03487, (2022).

[Con77] F. Constantinescu, *Nontriviality of the scattering matrix for weakly coupled ϕ_3^4 models*, Ann. Physics **108** (1977), no. 1, 37–48, DOI 10.1016/0003-4916(77)90349-9. MR452341

[CR75] P. Courrège and P. Renouard, *Oscillateur anharmonique mesures quasi-invariantes sur $C(\mathbf{R}, \mathbf{R})$ et théorie quantique des champs en dimension $d = 1$* (French), Astérisque, Nos. 22-23, Soc. Math. France, Paris, 1975, pp. 3–245. MR496165

[DE77] J. Dimock and J.-P. Eckmann, *Spectral properties and bound-state scattering for weakly coupled $\lambda P(\varphi)_2$ models*, Ann. Physics **103** (1977), no. 2, 289–314, DOI 10.1016/S0003-4916(97)90002-6. MR428957

[Dim74] J. Dimock, *Asymptotic perturbation expansion in the $P(\phi)_2$ quantum field theory*, Comm. Math. Phys. **35** (1974), 347–356. MR334756

[Dim77] J. Dimock, *The non-relativistic limit of $P(\phi)_2$ quantum field theories: two-particle phenomena*, Comm. Math. Phys. **57** (1977), no. 1, 51–66. MR475455

[Dim14] J. Dimock, *The renormalization group according to Balaban III. Convergence*, Ann. Henri Poincaré **15** (2014), no. 11, 2133–2175, DOI 10.1007/s00023-013-0303-3. MR3268826

[DM73] R. L. Dobrušin and R. A. Minlos, *Construction of a one-dimensional quantum field by means of a continuous Markov field* (Russian), Funkcional. Anal. i Priložen. **7** (1973), no. 4, 81–82. MR378654

[Don81] M. Donald, *The classical field limit of $P(\varphi)_2$ quantum field theory*, Comm. Math. Phys. **79** (1981), no. 2, 153–165. MR612245

[DPD03] G. Da Prato and A. Debussche, *Strong solutions to the stochastic quantization equations*, Ann. Probab. **31** (2003), no. 4, 1900–1916, DOI 10.1214/aop/1068646370. MR2016604

[DPD20] G. Da Prato and A. Debussche, *Gradient estimates and maximal dissipativity for the Kolmogorov operator in Φ_2^4*, Electron. Commun. Probab. **25** (2020), Paper No. 9, 16, DOI 10.1214/20-ecp294. MR4066302

[DPT00] G. Da Prato and L. Tubaro, *Self-adjointness of some infinite-dimensional elliptic operators and application to stochastic quantization*, Probab. Theory Related Fields **118** (2000), no. 1, 131–145, DOI 10.1007/PL00008739. MR1785456

[DVG21] F. C. De Vecchi and M. Gubinelli, *A note on supersymmetry and stochastic differential equations*, Geometry and invariance in stochastic dynamics, Springer Proc. Math. Stat., vol. 378, Springer, Cham, [2021] ©2021, pp. 71–87, DOI 10.1007/978-3-030-87432-2_5. MR4432535

[Ebe99] A. Eberle, *Uniqueness and non-uniqueness of semigroups generated by singular diffusion operators*, Lecture Notes in Mathematics, vol. 1718, Springer-Verlag, Berlin, 1999, DOI 10.1007/BFb0103045. MR1734956

[Eck76] J.-P. Eckmann, *Asymptotic perturbation expansion for the S-matrix in $P(\phi)_2$ quantum field theory models*, Quantum dynamics: models and mathematics (Proc. Sympos., Centre for Interdisciplinary Res., Bielefeld Univ., Bielefeld, 1975), Springer, Vienna-New York, 1976, pp. 59–68. MR673614

[EE79] J.-P. Eckmann and H. Epstein, *Time-ordered products and Schwinger functions*, Comm. Math. Phys. **64** (1978/79), no. 2, 95–130. MR519920

[EEF77] J.-P. Eckmann, H. Epstein, and J. Fröhlich, *Asymptotic perturbation expansion for the S-matrix and the definition of time ordered functions in relativistic quantum field models*, Ann. Inst. H. Poincaré Sect. A (N.S.) **25** (1976/77), no. 1, 1–34. MR418716

[EMS75] J.-P. Eckmann, J. Magnen, and R. Sénéor, *Decay properties and Borel summability for the Schwinger functions in $P(\phi)_2$ theories*, Comm. Math. Phys. **39** (1974/75), 251–271. MR366266

[EO71] J.-P. Eckmann and K. Osterwalder, *On the uniqueness of the Hamiltonian and of the representation of the* CCR *for the quartic boson interaction in three dimensions*, Helv. Phys. Acta **44** (1971), 884–909. MR359612

[Fab70] J. D. Fabrey, *Exponential representations of the canonical commutation relations*, Comm. Math. Phys. **19** (1970), 1–30. MR272311

[Fed71] P. Federbush, *Unitary renormalization of* $[\phi^4]_{2+1}$, Comm. Math. Phys. **21** (1971), 261–268. MR292432

[Fel74] J. Feldman, *The* $\lambda\varphi_3^4$ *field theory in a finite volume*, Comm. Math. Phys. **37** (1974), 93–120. MR384003

[FFS92] R. Fernández, J. Fröhlich, and A. D. Sokal, *Random walks, critical phenomena, and triviality in quantum field theory*, Texts and Monographs in Physics, Springer-Verlag, Berlin, 1992, DOI 10.1007/978-3-662-02866-7. MR1219313

[FG19] M. Furlan and M. Gubinelli, *Weak universality for a class of 3d stochastic reaction-diffusion models*, Probab. Theory Related Fields **173** (2019), no. 3-4, 1099–1164, DOI 10.1007/s00440-018-0849-6. MR3936152

[FHSX21] T. Funaki, M. Hoshino, S. Sethuraman, and B. Xie, *Asymptotics of PDE in random environment by paracontrolled calculus* (English, with English and French summaries), Ann. Inst. Henri Poincaré Probab. Stat. **57** (2021), no. 3, 1702–1735, DOI 10.1214/20-aihp1129. MR4291459

[FO76] J. S. Feldman and K. Osterwalder, *The Wightman axioms and the mass gap for weakly coupled* $(\Phi^4)_3$ *quantum field theories*, Ann. Physics **97** (1976), no. 1, 80–135, DOI 10.1016/0003-4916(76)90223-2. MR416337

[FOT94] M. Fukushima, Y. Ōshima, and M. Takeda, *Dirichlet forms and symmetric Markov processes*, De Gruyter Studies in Mathematics, vol. 19, Walter de Gruyter & Co., Berlin, 1994, DOI 10.1515/9783110889741. MR1303354

[FP77] J. Fröhlich and Y. M. Park, *Remarks on exponential interactions and the quantum sine-Gordon equation in two space-time dimensions*, Helv. Phys. Acta **50** (1977), no. 3, 315–329. MR456220

[FR96] J. Fritz and B. Rüdiger, *Approximation of a one-dimensional stochastic PDE by local mean field type lattice systems*, Nonlinear stochastic PDEs (Minneapolis, MN, 1994), IMA Vol. Math. Appl., vol. 77, Springer, New York, 1996, pp. 111–125, DOI 10.1007/978-1-4613-8468-7_6. MR1395895

[FRa77] J. S. Feldman and R. Rączka, *The relativistic field equation of the* $\lambda\Phi_3^4$ *quantum field theory*, Ann. Physics **108** (1977), no. 1, 212–229, DOI 10.1016/0003-4916(77)90357-8. MR452279

[Frö74] J. Fröhlich, *Verification of axioms for Euclidean and relativistic fields and Haag's theorem in a class of* $P(\varphi)_2$*-models*, Ann. Inst. H. Poincaré Sect. A (N.S.) **21** (1974), 271–317. MR373494

[FS76] J. Fröhlich and E. Seiler, *The massive Thirring-Schwinger model (QED2): convergence of perturbation theory and particle structure*, Helv. Phys. Acta **49** (1976), no. 6, 889–924. MR443693

[FS77] J. Fröhlich and B. Simon, *Pure states for general* $P(\phi)_2$ *theories: construction, regularity and variational equality*, Ann. of Math. (2) **105** (1977), no. 3, 493–526, DOI 10.2307/1970921. MR436829

[FSS76] J. Fröhlich, B. Simon, and T. Spencer, *Infrared bounds, phase transitions and continuous symmetry breaking*, Comm. Math. Phys. **50** (1976), no. 1, 79–95. MR421531

[Fun83] T. Funaki, *Random motion of strings and related stochastic evolution equations*, Nagoya Math. J. **89** (1983), 129–193, DOI 10.1017/S0027763000020298. MR692348

[GH19] M. Gubinelli and M. Hofmanová, *Global solutions to elliptic and parabolic Φ^4 models in Euclidean space*, Comm. Math. Phys. **368** (2019), no. 3, 1201–1266, DOI 10.1007/s00220-019-03398-4. MR3951704

[GH21] M. Gubinelli and M. Hofmanová, *A PDE construction of the Euclidean ϕ_3^4 quantum field theory*, Comm. Math. Phys. **384** (2021), no. 1, 1–75, DOI 10.1007/s00220-021-04022-0. MR4252872

[Gie83] R. Gielerak, *Verification of the global Markov property in some class of strongly coupled exponential interactions*, J. Math. Phys. **24** (1983), no. 2, 347–355, DOI 10.1063/1.525686. MR692312

[GIP15] M. Gubinelli, P. Imkeller, and N. Perkowski, *Paracontrolled distributions and singular PDEs*, Forum Math. Pi **3** (2015), e6, 75, DOI 10.1017/fmp.2015.2. MR3406823

[GJ68] J. Glimm and A. Jaffe, *A $\lambda\phi^4$ quantum field without cutoffs. I*, Phys. Rev. (2) **176** (1968), 1945–1951. MR247845

[GJ70a] J. Glimm and A. Jaffe, *Rigorous quantum field theory models*, Bull. Amer. Math. Soc. **76** (1970), 407–410, DOI 10.1090/S0002-9904-1970-12492-2. MR253675

[GJ70b] J. Glimm and A. Jaffe, *The $\lambda(\phi^4)_2$ quantum field theory without cutoffs. III. The physical vacuum*, Acta Math. **125** (1970), 203–267, DOI 10.1007/BF02392335. MR269234

[GJ70c] J. Glimm and A. Jaffe, *The $\lambda(\varphi^4)_2$ quantum field theory without cutoffs. II. The field operators and the approximate vacuum*, Ann. of Math. (2) **91** (1970), 362–401, DOI 10.2307/1970582. MR256677

[GJ72] J. Glimm and A. Jaffe, *Boson quantum field models*, Mathematics of contemporary physics (Proc. Instructional Conf. (NATO Advanced Study Inst.), Bedford Coll., London, 1971), Academic Press, London-New York, 1972, pp. 77–143. MR674511

[GJ73] J. Glimm and A. Jaffe, *Positivity of the ϕ_3^4 Hamiltonian*, Fortschr. Physik **21** (1973), 327–376, DOI 10.1002/prop.19730210702. MR408581

[GJ81] J. Glimm and A. Jaffe, *Quantum physics*, Springer-Verlag, New York-Berlin, 1981. A functional integral point of view. MR0628000

[GJS74] J. Glimm, A. Jaffe, and T. Spencer, *The Wightman axioms and particle structure in the $\mathcal{P}(\phi)_2$ quantum field model*, Ann. of Math. (2) **100** (1974), 585–632, DOI 10.2307/1970959. MR363256

[GK86] K. Gawędzki and A. Kupiainen, *Asymptotic freedom beyond perturbation theory*, Phénomènes critiques, systèmes aléatoires, théories de jauge, Part I, II (Les Houches, 1984), North-Holland, Amsterdam, 1986, pp. 185–292. MR880526

[Gla74] V. Glaser, *On the equivalence of the Euclidean and Wightman formulation of field theory*, Comm. Math. Phys. **37** (1974), 257–272. MR353859

[Gli68] J. Glimm, *Boson fields with the $:\Phi^4:$ interaction in three dimensions*, Comm. Math. Phys. **10** (1968), 1–47. MR231601

[GLP99] G. Giacomin, J. L. Lebowitz, and E. Presutti, *Deterministic and stochastic hydrodynamic equations arising from simple microscopic model systems*, Stochastic partial differential equations: six perspectives, Math. Surveys Monogr., vol. 64, Amer. Math. Soc., Providence, RI, 1999, pp. 107–152, DOI 10.1090/surv/064/03. MR1661764

[GR73] F. Guerra and P. Ruggiero, *New interpretation of the Euclidean-Markov field in the framework of physical Minkowski space-time*, Phys. Rev. Lett. **31** (1973), 1022–1025.

[GRS72] F. Guerra, L. Rosen, and B. Simon, *Nelson's symmetry and the infinite volume behavior of the vacuum in $P(\phi)_2$*, Comm. Math. Phys. **27** (1972), 10–22. MR311242

[GRS75] F. Guerra, L. Rosen, and B. Simon, *The $\mathbf{P}(\phi)_2$ Euclidean quantum field theory as classical statistical mechanics. I, II*, Ann. of Math. (2) **101** (1975), 111–189; ibid. (2) 101 (1975), 191–259, DOI 10.2307/1970988. MR378670

[Gub04] M. Gubinelli, *Controlling rough paths*, J. Funct. Anal. **216** (2004), no. 1, 86–140, DOI 10.1016/j.jfa.2004.01.002. MR2091358

[Gub10] M. Gubinelli, *Ramification of rough paths*, J. Differential Equations **248** (2010), no. 4, 693–721, DOI 10.1016/j.jde.2009.11.015. MR2578445

[Gue72] F. Guerra, *Uniqueness of the vacuum energy density and van Hove phenomenon in the infinite-volume limit for two-dimensional self-coupled Bose fields*, Phys. Rev. Lett. **28** (1972), 1213–1215.

[Hai14] M. Hairer, *A theory of regularity structures*, Invent. Math. **198** (2014), no. 2, 269–504, DOI 10.1007/s00222-014-0505-4. MR3274562

[Heg74] G. C. Hegerfeldt, *From Euclidean to relativistic fields and on the notion of Markoff fields*, Comm. Math. Phys. **35** (1974), 155–171. MR349175

[Hep69] K. Hepp, *Théorie de la renormalisation* (French), Lecture Notes in Physics, Vol. 2, Springer-Verlag, Berlin-New York, 1969. Cours donné à l'École Polytechnique, Paris. MR277208

[HI18] M. Hairer and M. Iberti, *Tightness of the Ising-Kac model on the two-dimensional torus*, J. Stat. Phys. **171** (2018), no. 4, 632–655, DOI 10.1007/s10955-018-2033-x. MR3790155

[HK71] R. Høegh-Krohn, *A general class of quantum fields without cut-offs in two space-time dimensions*, Comm. Math. Phys. **21** (1971), 244–255. MR292433

[HKK20] M. Hoshino, H. Kawabi, and S. Kusuoka, *Stochastic quantization associated with the* $\exp(\Phi)_2$*-quantum field model driven by space-time white noise on the torus in the full* L^1*-regime*, Probab. Theory Related Fields **185** (2023), no. 1-2, 391–447, DOI 10.1007/s00440-022-01126-z. MR4528973

[HKK21] M. Hoshino, H. Kawabi, and S. Kusuoka, *Stochastic quantization associated with the* $\exp(\Phi)_2$*-quantum field model driven by space-time white noise on the torus*, J. Evol. Equ. **21** (2021), no. 1, 339–375, DOI 10.1007/s00028-020-00583-0. MR4238209

[HKPS93] T. Hida, H.-H. Kuo, J. Potthoff, and L. Streit, *White noise*, Mathematics and its Applications, vol. 253, Kluwer Academic Publishers Group, Dordrecht, 1993. An infinite-dimensional calculus, DOI 10.1007/978-94-017-3680-0. MR1244577

[HL18] M. Hairer and C. Labbé, *Multiplicative stochastic heat equations on the whole space*, J. Eur. Math. Soc. (JEMS) **20** (2018), no. 4, 1005–1054, DOI 10.4171/JEMS/781. MR3779690

[HM18a] M. Hairer and K. Matetski, *Discretisations of rough stochastic PDEs*, Ann. Probab. **46** (2018), no. 3, 1651–1709, DOI 10.1214/17-AOP1212. MR3785597

[HM18b] M. Hairer and J. Mattingly, *The strong Feller property for singular stochastic PDEs* (English, with English and French summaries), Ann. Inst. Henri Poincaré Probab. Stat. **54** (2018), no. 3, 1314–1340, DOI 10.1214/17-AIHP840. MR3825883

[HS16] M. Hairer and H. Shen, *The dynamical sine-Gordon model*, Comm. Math. Phys. **341** (2016), no. 3, 933–989, DOI 10.1007/s00220-015-2525-3. MR3452276

[HS22] M. Hairer and R. Steele, *The* Φ^4_3 *measure has sub-Gaussian tails*, J. Stat. Phys. **186** (2022), no. 3, Paper No. 38, 25, DOI 10.1007/s10955-021-02866-3. MR4375843

[HX18] M. Hairer and W. Xu, *Large-scale behavior of three-dimensional continuous phase coexistence models*, Comm. Pure Appl. Math. **71** (2018), no. 4, 688–746, DOI 10.1002/cpa.21738. MR3772400

[IW89] N. Ikeda and S. Watanabe, *Stochastic differential equations and diffusion processes*, 2nd ed., North-Holland Mathematical Library, vol. 24, North-Holland Publishing Co., Amsterdam; Kodansha, Ltd., Tokyo, 1989. MR1011252

[Iwa87a] K. Iwata, *An infinite-dimensional stochastic differential equation with state space* $C(\mathbf{R})$, Probab. Theory Related Fields **74** (1987), no. 1, 141–159, DOI 10.1007/BF01845644. MR863723

[Iwa87b] K. Iwata, *Reversible measures of a* $P(\phi)_1$*-time evolution*, Probabilistic methods in mathematical physics (Katata/Kyoto, 1985), Academic Press, Boston, MA, 1987, pp. 195–209. MR933824

[Jan97] S. Janson, *Gaussian Hilbert spaces*, Cambridge Tracts in Mathematics, vol. 129, Cambridge University Press, Cambridge, 1997, DOI 10.1017/CBO9780511526169. MR1474726

[JLM85] G. Jona-Lasinio and P. K. Mitter, *On the stochastic quantization of field theory*, Comm. Math. Phys. **101** (1985), no. 3, 409–436. MR815192

[Jos65] R. Jost, *The general theory of quantized fields*, Lectures in Applied Mathematics, American Mathematical Society, Providence, RI, 1965. Mark Kac, editor. MR177667

[Kah85] J.-P. Kahane, *Sur le chaos multiplicatif* (French), Ann. Sci. Math. Québec **9** (1985), no. 2, 105–150. MR829798

[KFP83] A. Klein and J. Fernando Perez, *Supersymmetry and dimensional reduction: a non-perturbative proof*, Phys. Lett. B **125** (1983), no. 6, 473–475, DOI 10.1016/0370-2693(83)91329-1. MR768150

[Kle78] A. Klein, *The semigroup characterization of Osterwalder-Schrader path spaces and the construction of Euclidean fields*, J. Functional Analysis **27** (1978), no. 3, 277–291, DOI 10.1016/0022-1236(78)90009-5. MR496172

[KLP84] A. Klein, L. J. Landau, and J. F. Perez, *Supersymmetry and the Parisi-Sourlas dimensional reduction: a rigorous proof*, Comm. Math. Phys. **94** (1984), no. 4, 459–482. MR763747

[KR07] H. Kawabi and M. Röckner, *Essential self-adjointness of Dirichlet operators on a path space with Gibbs measures via an SPDE approach*, J. Funct. Anal. **242** (2007), no. 2, 486–518, DOI 10.1016/j.jfa.2006.06.017. MR2274818

[Kup16] A. Kupiainen, *Renormalization group and stochastic PDEs*, Ann. Henri Poincaré **17** (2016), no. 3, 497–535, DOI 10.1007/s00023-015-0408-y. MR3459120

[Kus92] S. Kusuoka, *Høegh-Krohn's model of quantum fields and the absolute continuity of measures*, Ideas and methods in quantum and statistical physics (Oslo, 1988), Cambridge Univ. Press, Cambridge, 1992, pp. 405–424. MR1190536

[Kus20] S. Kusuoka, *An improvement of the integrability of the state space of the Φ^4_3-process and the support of the Φ^4_3-measure constructed by the limit of stationary processes of approximating stochastic quantization equations*, Math. J. Okayama Univ. **65** (2023), 97–116, DOI 10.1017/s001708952200012x. MR4520573

[LN74] A. Lenard and C. M. Newman, *Infinite volume asymptotics in $P(\phi)_2$ field theory*, Comm. Math. Phys. **39** (1974), 243–250. MR356776

[Mar79] R. Marcus, *Stochastic diffusion on an unbounded domain*, Pacific J. Math. **84** (1979), no. 1, 143–153. MR559632

[Mil15] M. E. Miller, *The origins of Schwinger's Euclidean Green's functions*, Stud. Hist. Philos. Sci. B Stud. Hist. Philos. Modern Phys. **50** (2015), 5–12, DOI 10.1016/j.shpsb.2015.01.008. MR3343337

[Mit91] S. K. Mitter, *Markov random fields, stochastic quantization and image analysis*, Applied and industrial mathematics (Venice, 1989), Math. Appl., vol. 56, Kluwer Acad. Publ., Dordrecht, 1991, pp. 101–109, DOI 10.1007/978-94-009-1994-5_4. MR1147193

[MP19] J. Martin and N. Perkowski, *Paracontrolled distributions on Bravais lattices and weak universality of the 2d parabolic Anderson model* (English, with English and French summaries), Ann. Inst. Henri Poincaré Probab. Stat. **55** (2019), no. 4, 2058–2110, DOI 10.1214/18-AIHP942. MR4029148

[MR92] Z. M. Ma and M. Röckner, *Introduction to the theory of (nonsymmetric) Dirichlet forms*, Universitext, Springer-Verlag, Berlin, 1992, DOI 10.1007/978-3-642-77739-4. MR1214375

[MR99] R. Mikulevicius and B. L. Rozovskii, *Martingale problems for stochastic PDE's*, Stochastic partial differential equations: six perspectives, Math. Surveys Monogr., vol. 64, Amer. Math. Soc., Providence, RI, 1999, pp. 243–325, DOI 10.1090/surv/064/06. MR1661767

[MS76] J. Magnen and R. Sénéor, *The infinite volume limit of the ϕ^4_3 model*, Ann. Inst. H. Poincaré Sect. A (N.S.) **24** (1976), no. 2, 95–159, DOI 10.1007/s11245-005-1376-5. MR406217

[MS77] J. Magnen and R. Sénéor, *Phase space cell expansion and Borel summability for the Euclidean ϕ^4_3 theory*, Comm. Math. Phys. **56** (1977), no. 3, 237–276. MR523030

[MW17a] J.-C. Mourrat and H. Weber, *Convergence of the two-dimensional dynamic Ising-Kac model to Φ^4_2*, Comm. Pure Appl. Math. **70** (2017), no. 4, 717–812, DOI 10.1002/cpa.21655. MR3628883

[MW17b] J.-C. Mourrat and H. Weber, *The dynamic Φ^4_3 model comes down from infinity*, Comm. Math. Phys. **356** (2017), no. 3, 673–753, DOI 10.1007/s00220-017-2997-4. MR3719541

[MW17c] J.-C. Mourrat and H. Weber, *Global well-posedness of the dynamic Φ^4 model in the plane*, Ann. Probab. **45** (2017), no. 4, 2398–2476, DOI 10.1214/16-AOP1116. MR3693966

[MW20] A. Moinat and H. Weber, *Space-time localisation for the dynamic Φ^4_3 model*, Comm. Pure Appl. Math. **73** (2020), no. 12, 2519–2555, DOI 10.1002/cpa.21925. MR4164267

[Nel66] E. Nelson, *A quartic interaction in two dimensions*, Mathematical Theory of Elementary Particles (Proc. Conf., Dedham, Mass., 1965), MIT Press, Cambridge, Mass.-London, 1966, pp. 69–73. MR210416

[Nel73a] E. Nelson, *Construction of quantum fields from Markoff fields*, J. Functional Analysis **12** (1973), 97–112, DOI 10.1016/0022-1236(73)90091-8. MR343815

[Nel73b] E. Nelson, *The free Markoff field*, J. Functional Analysis **12** (1973), 211–227, DOI 10.1016/0022-1236(73)90025-6. MR343816

[Nel73c] E. Nelson, *Probability theory and Euclidean field theory*, Constructive Quantum Field Theory (G. Velo and A. Wightman, eds.), Lecture Notes in Physics, vol. 25, Springer, Berlin, Heidelberg, 1973, pp. 94–124.

[Nel73d] E. Nelson, *Quantum fields and Markoff fields*, Partial differential equations (Proc. Sympos. Pure Math., Vol. XXIII, Univ. California, Berkeley, Calif., 1971), Proc. Sympos. Pure Math., Vol. XXIII, Amer. Math. Soc., Providence, RI, 1973, pp. 413–420. MR337206

[Nel75] E. Nelson, *Remarks on Markov field equations*, Functional integration and its applications (Proc. Internat. Conf., London, 1974), Clarendon Press, Oxford, 1975, pp. 136–143. MR484174

[Nel81] N. Wiener, *Collected works. Vol. I*, Mathematicians of Our Time, vol. 10, MIT Press, Cambridge, Mass.-London, 1976. Mathematical philosophy and foundations; potential theory; Brownian movement, Wiener integrals, ergodic and chaos theories, turbulence and statistical mechanics; With commentaries; Edited by P. Masani. MR532698

[NP12] I. Nourdin and G. Peccati, *Normal approximations with Malliavin calculus*, Cambridge Tracts in Mathematics, vol. 192, Cambridge University Press, Cambridge, 2012. From Stein's method to universality, DOI 10.1017/CBO9781139084659. MR2962301

[ORSW20] T. Oh, T. Robert, P. Sosoe, and Y. Wang, *Invariant Gibbs dynamics for the dynamical sine-Gordon model*, Proc. Roy. Soc. Edinburgh Sect. A **151** (2021), no. 5, 1450–1466, DOI 10.1017/prm.2020.68. MR4313569

[OS73] K. Osterwalder and R. Schrader, *Axioms for Euclidean Green's functions*, Comm. Math. Phys. **31** (1973), 83–112. MR329492

[OS75] K. Osterwalder and R. Schrader, *Axioms for Euclidean Green's functions. II*, Comm. Math. Phys. **42** (1975), 281–305. With an appendix by Stephen Summers. MR376002

[OS76] K. Osterwalder and R. Sénéor, *The scattering matrix is non-trivial for weakly coupled $P(\phi)_2$ models*, Helv. Phys. Acta **49** (1976), no. 4, 525–535. MR411461

[Ost73] K. Osterwalder, *Euclidean Green's functions and Wightman distributions*, In: Velo G., Wightman A. (eds) Constructive Quantum Field Theory, Lecture Notes in Physics, vol. 25, Springer, Berlin, Heidelberg, 1973.

[Ost82] P. Diţă, V. Georgescu, and R. Purice (eds.), *Gauge theories: fundamental interactions and rigorous results*, Progress in Physics, vol. 5, Birkhäuser, Boston, MA, 1982. Lectures presented at the International Summer School on Theoretical Physics held in Poiana Brasov, August 25–September 7, 1981. MR704366

[Par76] Y. M. Park, *Uniform bounds of the pressures of the $\lambda\phi^4_3$ field model*, J. Mathematical Phys. **17** (1976), no. 6, 1073–1075, DOI 10.1063/1.522995. MR406222

[Par77] Y. M. Park, *Convergence of lattice approximations and infinite volume limit in the $(\lambda\phi^4 - \sigma\phi^2 - \tau\phi)_3$ field theory*, J. Mathematical Phys. **18** (1977), no. 3, 354–366, DOI 10.1063/1.523277. MR432062

[PS79] G. Parisi and N. Sourlas, *Random magnetic fields, supersymmetry, and negative dimensions*, Phys. Rev. Lett. **43** (1979), 744–745.

[PW81] G. Parisi and Y. S. Wu, *Perturbation theory without gauge fixing*, Sci. Sinica **24** (1981), no. 4, 483–496. MR626795

[Röc85] M. Röckner, *Generalized Markov fields and Dirichlet forms*, Acta Appl. Math. **3** (1985), no. 3, 285–311, DOI 10.1007/BF00047332. MR790552

[Röc88] M. Röckner, *Traces of harmonic functions and a new path space for the free quantum field*, J. Funct. Anal. **79** (1988), no. 1, 211–249, DOI 10.1016/0022-1236(88)90037-7. MR950091

[RZZ17a] M. Röckner, R. Zhu, and X. Zhu, *Ergodicity for the stochastic quantization problems on the 2D-torus*, Comm. Math. Phys. **352** (2017), no. 3, 1061–1090, DOI 10.1007/s00220-017-2865-2. MR3631399

[RZZ17b] M. Röckner, R. Zhu, and X. Zhu, *Restricted Markov uniqueness for the stochastic quantization of $P(\Phi)_2$ and its applications*, J. Funct. Anal. **272** (2017), no. 10, 4263–4303, DOI 10.1016/j.jfa.2017.01.023. MR3626040

[Sch71] S. Scheinberg, *Fatou's lemma in normed linear spaces*, Pacific J. Math. **38** (1971), 233–238. MR308757

[Sch94] S. S. Schweber, *QED and the men who made it: Dyson, Feynman, Schwinger, and Tomonaga*, Princeton Series in Physics, Princeton University Press, Princeton, NJ, 1994. MR1270676

[Shi95] I. Shigekawa, *An example of regular (r, p)-capacity and essential self-adjointness of a diffusion operator in infinite dimensions*, J. Math. Kyoto Univ. **35** (1995), no. 4, 639–651, DOI 10.1215/kjm/1250518653. MR1365253

[Sim74] B. Simon, *The $P(\phi)_2$ Euclidean (quantum) field theory*, Princeton Series in Physics, Princeton University Press, Princeton, NJ, 1974. MR489552

[Sim05] B. Simon, *Functional integration and quantum physics*, 2nd ed., AMS Chelsea Publishing, Providence, RI, 2005, DOI 10.1090/chel/351. MR2105995

[SS76] E. Seiler and B. Simon, *Nelson's symmetry and all that in the Yukawa2 and $(\phi^4)_3$ field theories*, Ann. Physics **97** (1976), no. 2, 470–518, DOI 10.1016/0003-4916(76)90044-0. MR438960

[Ste70] E. M. Stein, *Singular integrals and differentiability properties of functions*, Princeton Mathematical Series, No. 30, Princeton University Press, Princeton, NJ, 1970. MR290095

[Str81] R. F. Streater, *Euclidean quantum mechanics and stochastic integrals*, Stochastic integrals (Proc. Sympos., Univ. Durham, Durham, 1980), Lecture Notes in Math., vol. 851, Springer, Berlin, 1981, pp. 371–393. MR620996

[Str13] F. Strocchi, *An introduction to non-perturbative foundations of quantum field theory*, International Series of Monographs on Physics, vol. 158, Oxford University Press, Oxford, 2013, DOI 10.1093/acprof:oso/9780199671571.001.0001. MR3186038

[Sum80] S. J. Summers, *A new proof of the asymptotic nature of perturbation theory in $P(\varphi)_2$ models*, Helv. Phys. Acta **53** (1980), no. 1, 1–30. MR594327

[Sum12] Stephen J. Summers, *A perspective on constructive quantum field theory*, 2012, `arXiv:1203.3991`.

[SW78] R. F. Streater and A. S. Wightman, *PCT, spin and statistics, and all that*, 2nd ed., Mathematical Physics Monograph Series, Benjamin/Cummings Publishing Co., Inc., Reading, Mass.-London-Amsterdam, 1978. MR468904

[Sym66] K. Symanzik, *Euclidean quantum field theory. I. Equations for a scalar model*, J. Mathematical Phys. **7** (1966), 510–525, DOI 10.1063/1.1704960. MR187686

[Sym69] K. Symanzik, *Euclidean quantum field theory*, Varenna 1968, Proceedings Of The Physics School On Local Quantum Theory, Academic Press, New York, 1969, pp. 152–226.

[SZ76] T. Spencer and F. Zirilli, *Scattering states and bound states in $\lambda\mathcal{P}(\phi)_2$*, Comm. Math. Phys. **49** (1976), no. 1, 1–16. MR413888

[SZZ21] H. Shen, R. Zhu, and X. Zhu, *An SPDE approach to perturbation theory of Φ^4_2: asymptoticity and short distance behavior*, Ann. Appl. Probab. **33** (2023), no. 4, 2600–2642, DOI 10.1214/22-aap1873. MR4612651

[SZZ22] H. Shen, R. Zhu, and X. Zhu, *Large N limit of the $O(N)$ linear sigma model in 3D*, Comm. Math. Phys. **394** (2022), no. 3, 953–1009, DOI 10.1007/s00220-022-04414-w. MR4470243

[Tak85] M. Takeda, *On the uniqueness of Markovian selfadjoint extension of diffusion operators on infinite-dimensional spaces*, Osaka J. Math. **22** (1985), no. 4, 733–742. MR815442

[TW20] P. Tsatsoulis and H. Weber, *Exponential loss of memory for the 2-dimensional Allen-Cahn equation with small noise*, Probab. Theory Related Fields **177** (2020), no. 1-2, 257–322, DOI 10.1007/s00440-019-00945-x. MR4095014

[Wig56] A. S. Wightman, *Quantum field theory in terms of vacuum expectation values*, Phys. Rev. (2) **101** (1956), 860–866. MR83914

[Zeg84] B. Zegarliński, *Uniqueness and the global Markov property for Euclidean fields: the case of general exponential interaction*, Comm. Math. Phys. **96** (1984), no. 2, 195–221. MR768255

[Zin95] Y. M. Zinoviev, *Equivalence of Euclidean and Wightman field theories*, Comm. Math. Phys. **174** (1995), no. 1, 1–27. MR1372797

Editorial Information

To be published in the *Memoirs*, a paper must be correct, new, nontrivial, and significant. Further, it must be well written and of interest to a substantial number of mathematicians. Piecemeal results, such as an inconclusive step toward an unproved major theorem or a minor variation on a known result, are in general not acceptable for publication.

Papers appearing in *Memoirs* are generally at least 80 and not more than 200 published pages in length. Papers less than 80 or more than 200 published pages require the approval of the Managing Editor of the Transactions/Memoirs Editorial Board. Published pages are the same size as those generated in the style files provided for $\mathcal{AMS}$-LaTeX.

Information on the backlog for this journal can be found on the AMS website starting from `http://www.ams.org/memo`.

A Consent to Publish is required before we can begin processing your paper. After a paper is accepted for publication, the Providence office will send a Consent to Publish and Copyright Agreement to all authors of the paper. By submitting a paper to the *Memoirs*, authors certify that the results have not been submitted to nor are they under consideration for publication by another journal, conference proceedings, or similar publication.

Information for Authors

Memoirs is an author-prepared publication. Once formatted for print and on-line publication, articles will be published as is with the addition of AMS-prepared frontmatter and backmatter. Articles are not copyedited; however, confirmation copy will be sent to the authors.

Initial submission. The AMS uses Centralized Manuscript Processing for initial submissions. Authors should submit a PDF file using the Initial Manuscript Submission form found at `www.ams.org/submission/memo`, or send one copy of the manuscript to the following address: Centralized Manuscript Processing, MEMOIRS OF THE AMS, 201 Charles Street, Providence, RI 02904-2294 USA. If a paper copy is being forwarded to the AMS, indicate that it is for *Memoirs* and include the name of the corresponding author, contact information such as email address or mailing address, and the name of an appropriate Editor to review the paper (see the list of Editors below).

The paper must contain a *descriptive title* and an *abstract* that summarizes the article in language suitable for workers in the general field (algebra, analysis, etc.). The *descriptive title* should be short, but informative; useless or vague phrases such as "some remarks about" or "concerning" should be avoided. The *abstract* should be at least one complete sentence, and at most 300 words. Included with the footnotes to the paper should be the 2020 *Mathematics Subject Classification* representing the primary and secondary subjects of the article. The classifications are accessible from `www.ams.org/msc/`. The Mathematics Subject Classification footnote may be followed by a list of *key words and phrases* describing the subject matter of the article and taken from it. Journal abbreviations used in bibliographies are listed in the latest *Mathematical Reviews* annual index. The series abbreviations are also accessible from `www.ams.org/msnhtml/serials.pdf`. To help in preparing and verifying references, the AMS offers MR Lookup, a Reference Tool for Linking, at `www.ams.org/mrlookup/`.

Electronically prepared manuscripts. The AMS encourages electronically prepared manuscripts, with a strong preference for $\mathcal{AMS}$-LaTeX. To this end, the Society has prepared $\mathcal{AMS}$-LaTeX author packages for each AMS publication. Author packages include instructions for preparing electronic manuscripts, samples, and a style file that generates the particular design specifications of that publication series.

Authors may retrieve an author package for *Memoirs of the AMS* from `www.ams.org/journals/memo/memoauthorpac.html`. The *AMS Author Handbook* is available in PDF format from the author package link. The author package can also be obtained free of charge by sending email to `tech-support@ams.org` or from the Publication Division,

American Mathematical Society, 201 Charles St., Providence, RI 02904-2294, USA. When requesting an author package, please specify the publication in which your paper will appear. Please be sure to include your complete mailing address.

After acceptance. The source files for the final version of the electronic manuscript should be sent to the Providence office immediately after the paper has been accepted for publication. The author should also submit a PDF of the final version of the paper to the editor, who will forward a copy to the Providence office.

Accepted electronically prepared files can be submitted via the web at `www.ams.org/submit-book-journal/`, sent via FTP, or sent on CD to the Electronic Prepress Department, American Mathematical Society, 201 Charles Street, Providence, RI 02904-2294 USA. TeX source files and graphic files can be transferred over the Internet by FTP to the Internet node `ftp.ams.org` (130.44.1.100). When sending a manuscript electronically via CD, please be sure to include a message indicating that the paper is for the *Memoirs*.

Electronic graphics. Comprehensive instructions on preparing graphics are available at `www.ams.org/authors/journals.html`. A few of the major requirements are given here.

Submit files for graphics as EPS (Encapsulated PostScript) files. This includes graphics originated via a graphics application as well as scanned photographs or other computer-generated images. If this is not possible, TIFF files are acceptable as long as they can be opened in Adobe Photoshop or Illustrator.

Authors using graphics packages for the creation of electronic art should also avoid the use of any lines thinner than 0.5 points in width. Many graphics packages allow the user to specify a "hairline" for a very thin line. Hairlines often look acceptable when proofed on a typical laser printer. However, when produced on a high-resolution laser imagesetter, hairlines become nearly invisible and will be lost entirely in the final printing process.

Screens should be set to values between 15% and 85%. Screens which fall outside of this range are too light or too dark to print correctly. Variations of screens within a graphic should be no less than 10%.

Any graphics created in color will be rendered in grayscale for the printed version unless color printing is authorized by the Managing Editor and the Publisher. In general, color graphics will appear in color in the online version.

Inquiries. Any inquiries concerning a paper that has been accepted for publication should be sent to `memo-query@ams.org` or directly to the Electronic Prepress Department, American Mathematical Society, 201 Charles St., Providence, RI 02904-2294 USA.

Editors

The AMS uses Centralized Manuscript Processing for initial submissions to AMS journals. Authors should follow instructions listed on the Initial Submission page found at www.ams.org/tran/transubmit.html.

This journal is designed particularly for long research papers, normally at least 80 pages in length, and groups of cognate papers in pure and applied mathematics. Papers intended for publication in the *Memoirs* should be addressed to one of the following editors. The AMS uses Centralized Manuscript Processing for initial submissions to AMS journals. Authors should follow instructions listed on the Initial Submission page found at www.ams.org/memo/memosubmit.html

SELECTED PUBLISHED TITLES IN THIS SERIES

1553 **Andreas Greven, Frank den Hollander, and Margriet Oomen,** Spatial Populations with Seed-Bank: Renormalisation on the Hierarchical Group, 2025

1552 **Jason Atnip, Hiroki Sumi, and Mariusz Urbański,** The Dynamics and Geometry of Semi-Hyperbolic Rational Semigroups, 2025

1551 **Ronan J. Conlon and Alix Deruelle,** Steady Gradient Kähler-Ricci Solitons on Crepant Resolutions of Calabi-Yau Cones, 2025

1550 **Jussi Behrndt, Fritz Gesztesy, and Marius Mitrea,** Sharp Boundary Trace Theory and Schrödinger Operators on Bounded Lipschitz Domains, 2025

1549 **Valentina Grazian and Chris Parker,** Saturated Fusion Systems on p-Groups of Maximal Class, 2025

1548 **Rui Loja Fernandes and Ivan Struchiner,** The Global Solutions to a Cartan's Realization Problem, 2025

1547 **Satoshi Kondo and Seidai Yasuda,** Arithmetic Quotients of the Bruhat-Tits Building for Projective General Linear Groups in Positive Characteristic, 2025

1546 **Filippo Mazzoli, Andrea Seppi, and Andrea Tamburelli,** Para-HyperKähler Geometry of the Deformation Space of Maximal Globally Hyperbolic Anti-de Sitter Three-Manifolds, 2025

1545 **Tobias Lenz,** G-Global Homotopy Theory and Algebraic K-Theory, 2025

1544 **James Maynard,** Primes in Arithmetic Progressions to Large Moduli III: Uniform Residue Classes, 2025

1543 **James Maynard,** Primes in Arithmetic Progressions to Large Moduli II: Well-Factorable Estimates, 2025

1542 **James Maynard,** Primes in Arithmetic Progressions to Large Moduli I: Fixed Residue Classes, 2025

1541 **Benedikt Ahrens, Paige Randall North, Michael Shulman, and Dimitris Tsementzis,** The Univalence Principle, 2025

1540 **J. Peter May, Mona Merling, and Angélica M. Osorno,** Equivariant Infinite Loop Space Theory: The Space Level Story, 2025

1539 **Rémi Coulon, Rhiannon Dougall, Barbara Schapira, and Samuel Tapie,** Twisted Patterson-Sullivan Measures and Applications to Amenability and Coverings, 2025

1538 **Jeffrey S. Case,** The Bigraded Rumin Complex via Differential Forms, 2025

1537 **Michael Drmota, Clemens Müllner, and Lukas Spiegelhofer,** Primes as Sums of Fibonacci Numbers, 2025

1536 **Matthew Baker and Oliver Lorscheid,** Foundations of Matroids, Part 1: Matroids without Large Uniform Minors, 2025

1535 **Christian Stump, Hugh Thomas, and Nathan Williams,** Cataland: Why the Fuß?, 2025

1534 **Padraig Condon, Alberto Espuny Díaz, António Girão, Daniela Kühn, and Deryk Osthus,** Hamiltonicity of Random Subgraphs of the Hypercube, 2024

1533 **Peter Rudzis,** Rough Collisions, 2024

1532 **Xiang Yu,** On the Problem of Infinite Spin in Total Collisions of the Planar N-Body Problem, 2024

1531 **Dirk Basson, Florian Breuer, and Richard Pink,** Drinfeld Modular Forms of Arbitrary Rank, 2024

1530 **Jianbei An, Gerhard Hiss, and Frank Lübeck,** The Inductive Blockwise Alperin Weight Condition for the Chevalley Groups $F_4(q)$, 2024

For a complete list of titles in this series, visit the AMS Bookstore at **www.ams.org/bookstore/memoseries/**.